Edward Thomas Booth, E Neale

Rough Notes on the Birds

Observed During Twenty-Five Years

Edward Thomas Booth, E Neale

Rough Notes on the Birds
Observed During Twenty-Five Years

ISBN/EAN: 9783742862839

Manufactured in Europe, USA, Canada, Australia, Japa

Cover: Foto ©berggeist007 / pixelio.de

Manufactured and distributed by brebook publishing software
(www.brebook.com)

Edward Thomas Booth, E Neale

Rough Notes on the Birds

ROUGH NOTES

ON THE

BIRDS OBSERVED

DURING TWENTY-FIVE YEARS' SHOOTING AND COLLECTING

IN THE

BRITISH ISLANDS

BY

E. T. BOOTH.

WITH PLATES FROM DRAWINGS BY E. NEALE,

TAKEN FROM SPECIMENS IN THE AUTHOR'S POSSESSION.

VOLUME III.

LONDON:

PUBLISHED BY R. H. PORTER, 6 TENTERDEN STREET, W.

AND

MESSRS. DULAU & CO., SOHO SQUARE, W.

1881–1887.

CONTENTS OF VOL. III.

CONTENTS OF VOL. III.

LIST OF PLATES TO VOL. III.

GREY-LAG GOOSE.

ANSER FERUS.

As far as I have been able to ascertain from personal observation, the Grey-lag Goose is the only representative of the family that remains with us a resident in a wild state throughout the year. Numbers of these birds rear their young in the more remote parts of Ross-shire, Sutherland, and Caithness, as well as on some of the Western Islands. Though proverbially one of the wildest of fowl during winter, those that nest on our shores lay aside their shyness until the young are able to provide for themselves. While crossing the moors in summer, I have now and then seen an old gander leave the cover growing near some small loch, and with outstretched neck attack any dogs that happened to be in attendance, the female at the same time being heard endeavouring to draw the young into some place of security. Such temerity, however, is only witnessed when surprised in an open spot; as a rule, the brood is kept secluded in some scrubby undergrowth or about the outskirts of an impenetrable swamp during the day.

To thoroughly explore the haunts of these birds and watch their habits and actions while feeding at daybreak or at dusk, one must be prepared either to camp out or to be willing to put up with such humble accommodation as is to be procured in the district. Having obtained permission from the lessee of an extensive shooting, I made an expedition to an unfrequented portion of his territory where Geese and many species of sea-fowl nested in almost perfect security. This happy hunting-ground, a range of several miles of rough moorland interspersed with numerous lochs, stretched down to the wildest portion of the rock-bound coast that overlooks the stormy Minch. An abridged extract from my notes for 1868 will give some idea as to the nature of the surroundings among which this species passes the spring and summer, as well as the somewhat primitive style of living to be anticipated by those who visit this part of the country in quest of sport.

An early start was made from the inn at which we were stopping on the morning of the 28th of May, and after a drive of some twelve or fourteen miles a tramp of six more was necessary along a swampy track utterly impassable for our four-wheeled conveyance. Having engaged the services of four or five of the resident fishermen from some small hovels near the coast, our baggage, stock of blankets, eatables, and drinkables were transferred to their fish-creels, which, though imparting to the contents a powerful aroma of herring, proved admirably adapted for carrying heavy loads. The travelling being exceedingly heavy and the miles long, a couple of hours were expended in reaching the shealing that was to form our headquarters in this dreary region: during the whole of our journey we had been exposed to a drifting mist and an occasional heavy downfall of rain. On arriving at our destination the interior of the edifice was ascertained to be somewhat dilapidated, and the purity of the atmosphere was by no means improved from the fact, unfortunately only too apparent, that shepherds and their dogs had recently sought shelter in the building. Some braxy mutton had been left behind, and the filth scattered about needed immediate removal, as the sanitary state of affairs was far from satisfactory. The fishermen from forty to fifty grazing on the short grass on the lawn of a shooting-lodge in a lonely strath in Caithness. The ground was perfectly uninclosed towards the moors, and the non-breeding birds (possibly the young of the previous year) usually put in an appearance on this their favourite feeding-place in the gloaming. In East Lothian this species, in company with Bean-Geese, some years ago became so destructive to the young crops that it was necessary to employ herds with a gun to scare away the large flocks that alighted on the fields. The birds rested in immense bodies, either on the sands of the Firth or out on the water, usually breaking up into smaller parties when coming inland to feed; constant persecution by day having at length put them on the alert, they remained afloat till after dusk, and only attempted to make for their feeding-grounds when the herds had withdrawn. In Norfolk a few of these Geese are occasionally shot on Breydon flats in cold weather, and I have now and then observed them frequenting the hills adjoining some of the larger broads; they are not, however, so numerous in this part of the county as several other species. While steaming through Yarmouth roads in February 1882, an immense string, extending at least a quarter of a mile in length, were noticed making their way in a southerly direction; their line of flight was held at some elevation, but the conspicuous grey patch on the shoulder of their wings left no doubt as to their identity. Large flocks also pass along at sea a short distance off the Sussex coast, though few, unless in exceedingly severe weather, penetrate into the marshes. During the winters from 1858 to 1860 I frequently shot over Pevensey Level, but not more than three or four individuals of this species were obtained. Young birds occasionally reach the south of England as early as October; in 1882 one was shot in Shoreham harbour during the second week in the month. The soft parts were as follows:—Iris hazel, with a yellow ring round eye. Beak a livid yellow with a tinge of flesh-colour; nail dirty white. Legs and feet a livid flesh-tinge.

After shooting two or three Grey-lags or Pink-footed Geese* on the fens near the Cam one wintry day early in December 1861, I was informed by the marshman† who accompanied me that in the recollection of either his father or grandfather Wild Geese had bred in the neighbourhood. If any nested in the locality in a wild state they belonged to this species; the low swampy islands with stunted coverts of alder, elder, and willow, on which I have occasionally seen their nests in the Highlands, correspond with certain parts of the unreclaimed slades and wastes that existed even in my own time in the fens about Waterbeach and Wicken.

On two or three occasions I have met with broods, reared from eggs taken in the district, feeding round farm-buildings in the Northern Highlands, and in one instance remarked the eagerness with which they responded to the call of the lassie who attended to their wants. Possibly their natural wildness would show itself in time, as statements have appeared in print to the effect that birds bred in confinement, if possessed of the full use of their wings, would not permit a near approach, though returning occasionally to the quarters in which they were reared.

* I am unable to lay my hands on the journal referring to this incident.

† A well-known character in the district, who rented a small extent of fen-land and was never without a trusty setter. Old John, with whom I enjoyed many a good day's sport, remembered well the time when the fens were free and unreclaimed; and though he still knew where the Snipes or fowl would drop when a flight arrived, was often unmindful of the changes that had taken place in the ownership of the land.

BEAN-GOOSE.

ANSER SEGETUM.

Though it has been asserted by several writers that this species breeds on the British Islands, I have been unable to discover any evidence tending to confirm the statement. In every instance where I visited the moors in the remote districts in the Northern Highlands on which these Geese were said to have nested, the birds invariably proved to be Grey-lags. A gamekeeper who looked after a wild stretch of moorland on the borders of Caithness and Sutherland, declared that several pairs remained during the summer in the vicinity of some of the larger lochs in both counties, and the man resolutely refused to be convinced as to the species even after I had carefully examined every bird he pointed out through the glasses. During my residence in East Lothian, Bean-Geese frequented the shores of the Firth of Forth about Aberlady Bay, and also at feeding-time proceeded inland on to the cultivated ground near Dirleton, Gullane, and North Berwick. The large flocks of these birds and Grey-lags proved so destructive to the young corn, that it became necessary to employ herds to keep them from the crops. They take but little notice of the labourers while at work; but, though all appear intent on searching for food, the slightest sign of danger is certain to attract the attention of the sentinel on duty.

Early in January 1864, while living in that district, I was proceeding one evening to dine with a neighbouring farmer when, as I imagined in the gloaming, a large flock of sheep were seen advancing over a field of young corn. Well aware that they had no business there, I went cautiously round at the back of the hedge to learn, if possible, where they were breaking through, and on looking over was surprised to discover the flock of sheep transformed into about five hundred wild Geese. I was totally unprepared at the moment to fire a heavy ten-bore breech-loader I carried; but before they got out of range I succeeded in bringing down two and wounding another, which was captured alive by a sheep-dog the following day. Had I known the birds were there, at least ten or a dozen must have been bagged, as they rose on wing so closely packed. The two killed were both of this species, though the third, which I did not see myself, appeared, from the description given by the shepherd into whose hands it fell, to be a Grey-lag Goose. I have seen these birds particularly numerous during some winters in the grass-marshes in the east of Norfolk; in this locality, the water-dykes being sufficiently wide to be navigable by a small punt, excellent sport was occasionally obtained. I noticed that the birds, if feeding out of range in the centre of some of the pieces, might now and then be driven by men stationed on the marsh-wall; they soon, however, became aware of this stratagem and sought safety in flight on the first signs of danger. When the weather remains open and the supply of food is plentiful, they often attain a great weight—a dozen I bagged on the 27th of December, 1871, on the Holmes Marshes* near Heigham Sounds averaging over eight and a half pounds each, the heaviest gander just turning the

* The Holmes Marshes were one of the most noted coursing-grounds in the east of Norfolk.

scale at 9$\frac{3}{4}$ lbs. On their first arrival in Norfolk towards the close of autumn, I have met with many far lighter birds; at that season they frequently collect in considerable numbers on the hills immediately adjoining the broads and offer fair chances for a punt-gun. I noted down in my journal that eight secured at one shot some years back only averaged six and a half pounds.

WHITE-FRONTED GOOSE.

ANSER ALBIFRONS.

Though I have met with White-fronted Geese in various parts of the country, from north to south, my notes contain little information about these handsome birds beyond references to the numbers slain, and I am able to add little or nothing to the knowledge already possessed concerning their habits.

This species is a visitor to all parts of the British Islands within a few miles of the coast-line, grass-marshes and the sandy flats along the shore, as well as saltwater mud-banks, being its favourite resorts. Some five-and-twenty years ago these Geese were exceedingly numerous during winter about the extensive feeding-grounds in Pevensey Level in Sussex, flocks of from two to three hundred being occasionally seen. Twice one morning in December 1859 I succeeded in stopping five birds with the two barrels of a heavy ten-bore, the river-banks on the first occasion, and the shelter afforded by a marsh-gate together with the adjoining posts and rails on the second, enabling a near approach to be made. These Geese have gradually deserted this famous wildfowl resort of former days; the last time I was in the district, some ten or twelve years ago, a great change was ascertained to have taken place, one of the local gunners (a "looker," who had accompanied me on my rounds since I first shot there) stating that he had not observed a flock of these birds for many winters. On the hills around Hickling Broad and on the Holmes marshes in the east of Norfolk they frequently consort with Bean-Geese, and the two species have been repeatedly killed at one discharge of either the punt- or shoulder-gun.

The young birds make their way as far south as the pools in the vicinity of the Land's End by the latter end of October: in 1880 I examined several killed within a few miles of Penzance. They also occasionally visit the Sussex coast at an early date: for several days during the second week in October 1883 about half a dozen wild Geese were observed flying about Shoreham, and one, which proved to be a juvenile of this species, was shot at a pool of brackish water between the harbour and Lancing on the 15th of the month. Although the adults are considered a dainty dish when properly served up, they by no means compare with the young, which certainly are, as far as I have been able to judge, the finest flavoured of any wildfowl to be met with on our shores. In the plumage of the first autumn there is a total absence of white marking at the base of the bill, and in that state they might readily be mistaken for immature Bean-Geese or Grey-lags, by those unacquainted with the colouring of the legs and beaks of each species. The tints of the soft parts of a young bird taken down shortly after it had been shot were as follows:—Upper and lower mandibles chrome-yellow, though not so deep or rich as in the adult, the nail a pale brownish tinge. Legs and toes the same tint of yellow as the mandibles, with webs slightly darker, nails pale slate-colour. The weight of the young one from which the colours were taken was within three ounces of four pounds.

The call-note of the White-fronted Goose somewhat resembles a hoarse peal of laughter, and the bird is occasionally styled the Laughing Goose.

BRENT GOOSE.

BERNICLA BRENTA.

To all parts of our southern and eastern coasts, where the shores are flat, the Brent Goose is a regular winter visitor; the numbers, however, that put in an appearance vary considerably. On the lochs and open sands of the Western Highlands I have met with a few of these birds, but never having followed them for sport in that quarter of Great Britain, can state little concerning their habits in such localities.

When living on the east coast of Ross-shire, I remarked that the Geese might usually be looked for towards the end of September, large numbers arriving during the first and second weeks in October. On the Dornoch and Cromarty Firths they would remain for a time till forced by continued persecution or the severity of the weather to work further south. If any faith can be put in the stories of the natives, they collected formerly in thousands on the Cromarty Firth in the neighbourhood of Invergordon. The increase in the numbers of shooters has, however, of late years rendered these favourite resorts unsafe, and the majority now pass south without halting for any length of time about their old quarters. I notice that one or two writers profess to doubt the statements of the old fowlers as to the numbers of Geese gathering in these parts; judging, however, from the immense flocks (acres of sandy muds and water at times being densely packed by birds) that came under my own observation eighteen or twenty years ago, there appears no reasonable cause for disbelief. While on their way towards the far north, Geese again make a descent in this locality as spring approaches, stragglers often remaining till late in April. On one occasion, in the middle of May 1869, I noticed at least twenty or thirty birds sunning themselves on a mud-bank in the Cromarty Firth between Foulis and Alness. The date at which Brent Geese may be looked for on the Norfolk muds and the flats off the south-eastern counties depends greatly on the weather.

Along the coasts of Kent, Sussex, and Hampshire, the bays, harbours, and estuaries are still visited in hard winters by large though gradually decreasing numbers. A short conversation with any of the antiquated native gunners would speedily convince the inquirer that in the opinion of these worthies the good old times have passed, never to return. So lately as 1878 and 1879, I have, however, watched the Black Geese (as Brents are known in this part) making their way along the Sussex coast to the west of Shoreham in bodies of two or three hundred, one flock following another in rapid succession for several hours. On the breaking-up of the frost the birds may again be seen from the shore as they wing their course towards the east. The experience of the past winter and the constant persecution to which they have been exposed have now rendered the survivors almost unapproachable, and it is seldom they offer chances to the persevering gunners.

Occasionally after protracted gales the unfortunate birds are thoroughly worn out. Early in 1870, while out in the Channel off Shoreham, I fell in with about a couple of hundred sleeping quietly on the water, utterly regardless of danger. The dull light of a dreary winter's day was fast closing in, and though the wind had dropped and the sea fallen, a long swell still rolled slowly in from the west. Suddenly a black line of fowl was sighted through the haze a short distance to the south of our course, and under the impression, at the first glance, that the birds were Scoters *, the head of the boat was turned in order to inspect their ranks †. It was not till we were within twenty yards that a single head was raised and the species ascertained. After paddling a yard or two, the whole body got on wing; so slow, however, were their movements, that ample time was given to make use effectively of each of the four barrels of two heavy 10-bores. Considering the weather to which they had been lately exposed, it was somewhat strange that the six birds secured were in fairly good condition—in fact above the average weight.

A green weed or grass, *Zostera marina* (slimy in texture and almost transparent), that flourishes only on the saltwater mud-banks, is the principal food of this species; small quantities of other marine vegetation are also consumed. Although a few Brent Geese have come under my observation on the broads of the eastern counties and on other pieces of inland water, I am of opinion that it is seldom, unless wounded, these birds quit salt water. According to the accounts of several of the gunners in the neighbourhood of Hickling, a large body of over one hundred of these Geese settled on the broad under the shelter of one of the hills during a heavy snow-storm some ten or a dozen years ago. One of the men worked his punt within range; but owing to the effects of the snow, it was impossible to discharge the gun, and in the end the birds escaped unmolested. In Pevensey Level, during severe winters, when large flocks had been noticed in the Channel, I have known one or two taken in traps set for Snipe or Teal in the open drains or grips—indicating that a few at least occasionally resorted to the marshes in that district in quest of food.

Numberless pages have been written by various authors describing the means and appliances by which the cunning of this wily species is supposed to be successfully baffled; the times and seasons most favourable for the prosecution of this somewhat uncertain pastime have also been fully given. More anxious, as a rule, to gain an insight into the manners and customs of fowl than to cause unnecessary slaughter, I am unable to record any shots worthy of notice that have fallen to my share. Large numbers of Brents have frequently been killed at one discharge, from forty to fifty, and even more, being occasionally picked up. The heaviest bag (well authenticated) that has come to my knowledge was obtained about the firths on the eastern coasts of Ross and Cromarty, some years ago, before big guns had become so plentiful, eighteen hundred Geese having been secured by a sportsman shooting in a double-handed punt in the space of six weeks. Geese, I learned from an old punt-gunner, were the sole objects of this fowler's ambition; on one occasion, having stopped at least thirty or forty Dun-birds accidentally in the dark, my informant declared that, on discovering the error, this eccentric individual allowed not a fowl to be recovered, leaving the result of the shot to the mercy of the tide.

In addition to the larger firths on the north-east coast of the Highlands, there were a few small muddy harbours and estuaries to which Geese made their way when driven from more attractive feeding-grounds. The Little Ferry near Golspie, owing to the absence of local gunners, usually afforded a safe resting-place to parties numbering from fifty to one hundred birds, and here I occasionally met with first-rate sport.

While dropping down the Dornoch Firth with the ebb-tide in a small single punt early one foggy morning, in the autumn of 1869, to ascertain if any Geese were harbouring near the bar, I passed several fishing-craft from Banf and Buckie, making for the Tain Sands to procure cargoes of mussels ‡. Pulling up alongside

* The Geese happened to be right over a bank often resorted to as a feeding-ground by both Common and Velvet Scoters.

† It is seldom in this part of the Channel, so exposed to south-west breezes, that a gunning-punt can be launched with hopes of success. For shooting- and collecting-purposes I usually went afloat in a boat of about sixteen feet in length, built for sailing or rowing fairly well when pulled by a couple of sturdy oarsmen. To deaden all sound the oars were muffled by sheep-skin, and the light-painted craft gliding quietly through the water, Divers, Grebes, and even fowl were often approached within range.

‡ Mussels used to be farmed on the Tain shore of the Dornoch Firth, large beds or "scarps" having been laid down. There was, in those days, a great demand for bait for the long-line haddock fishery, and numbers of boats from the east coast entered the firth for cargoes.

of a boat that had already taken the ground, I inquired of one of the hands who was watching the punt if they had noticed many Geese near the entrance to the firth. There was little in the way of fowl, I was told, to tempt one so far, but about the sand-banks there were numbers of fine seals. The curiosity of the whole of the crew appeared to be aroused by my reply that fowl alone were the objects of which I was in search, and two or three more turned out to view the stranger. "Eh, mon, but one's a grand beastie," added another; "he's woorth a poond." "An' ye pay for yer ain poother an' hail?"* was the next query from a solemn-looking old Highlander in a fantastic blue bonnet. On my stating that such was the case, they all turned in again, the last remarking, as he disappeared, "Eh, but ye're nae puir mon." The canny Highlanders could scarcely believe it possible that the "beastie woorth a poond" would not be preferred to a few Geese.

After a residence of a year or two on the coasts of the northern and eastern firths, I learned that seals were held in great estimation by the natives on account of the oil, which they firmly believed to possess miraculous powers for curing all manner of diseases. In order to satisfy a few poor old bodies who had expressed a wish for a small quantity of this healing, strengthening, and sight-giving remedy, I shot several seals (in those days by no means difficult to obtain), and cutting up the blubber into chunks, extracted the oil by boiling. The quality of the oil given away was reported to be far superior to any previously obtained in the neighbourhood, and the news spread far and wide. Very shortly empty bottles arrived, frequently from a distance, accompanied, in many instances, by a half-crown, with a humble request that the bottle might be filled. After a time a roaring trade might have been carried on, when the natives ascertained that the full bottle together with the half-crown were invariably returned.

* Shot is known among many of the natives of this district as "hail."

WHOOPER.

CYGNUS MUSICUS.

It is useless to enumerate every locality in which this species has come under my observation; in smaller or larger "herds"* these fine birds have been met with during winter along many portions of the open coast-line from Sutherland to Sussex. Though stragglers may not unfrequently be seen early in the season, it is usually a few days in advance of heavy gales of wind from the north or east or the setting in of frost and snow that Wild Swans put in an appearance in any numbers on the saltwater lochs and friths of the Northern Highlands—arriving somewhat later on the broads and meres of the eastern counties, whence they are shortly driven to the estuaries and mud-flats further south.

Whoopers occasionally penetrate long distances inland; when living in the west of Perthshire I ascertained that a pair of these birds took up their quarters during winter for three successive years on some small lochs near the head of Glenlyon. Their haunts in this wild and desolate glen, shut in by steep and rugged hills, were seldom intruded on; and they took little or no notice of any keepers or shepherds making their way across country by an old hill-track that ran along the loch-side. In December 1865 I went up the glen to learn whether the birds were Whoopers or Bewick's Swans, and on my first visit had not the slightest difficulty in perceiving that they belonged to the former species. On our approach the pair were dressing their feathers (one occasionally preening the neck of the other or bowing its head with a low chattering note) on the shore of the loch near the road. Not needing specimens, I made no attempt to obtain a shot, though both could doubtless have been procured with the greatest ease, the large slabs of rock and stone encumbering the track offering every chance for a successful stalk. When at length aroused they merely paddled off to the distance of eighty or one hundred yards, and, turning round, quietly regarded without the slightest symptoms of alarm the intruders on their domain. The assistance of the glasses was scarcely needed to establish their identity: the birds, judging by their comparative sizes and actions, were evidently male and female, the former weighing probably as much as four or five and twenty pounds, while the latter was considerably lighter. In 1866 I did not remain sufficiently long in the north to inspect the Swans, who were somewhat late in reaching the glen; the following year the snow fell early, and the pair having been reported at their accustomed quarters, the first opportunity was taken to again interview my old friends. On arriving in sight of the loch a heavy snow-squall was passing over, and for a time the birds remained undetected; at last, when the storm cleared off, they came in view, swimming slowly out of a small creek in which they had evidently sought shelter from the cutting blasts, the first glimpse being sufficient to show that they corresponded in every particular with the pair so closely examined on the former occasion. The wind now increasing rapidly and causing the snow to drift, rendered it necessary to beat a speedy retreat; so without delaying to make

* To the general reader many of the names bestowed on flocks of Wildfowl and Plover by those who claim to be professors of the art of fowling could not fail to be confusing; in these pages it is advisable to use only such terms as are perfectly intelligible. The meaning of a "herd" of Swans must be plain to all; but such expressions as a "little knob" of Teal or a "dopping" of Sheldrakes might prove perplexing.

further observations, we turned homewards down the glen. Before quitting the water-side the birds commenced to show signs of restlessness, and rising on wing made off towards another small loch, from which, however, they shortly returned, flying low in the face of the wind. On making inquiries I learned that the Swans deserted the loch this year at a much earlier date than usual; a severe frost having set in, it is probable that the surface would have been frozen over to such an extent as to interfere with their supply of food.

Though numberless chances of obtaining shots at these birds were lost through endeavours to gain an insight into their general habits and manner of feeding, I am able to add little or nothing to the knowledge already possessed concerning the species. Weeds dragged up from the bottom of freshwater lochs in the north, or the larger broads and pools in the south, appear to form a considerable portion of their diet; I also watched them on many occasions grazing on the grass-marshes and round the edges of the dykes in flat districts. In shallow bays and in the channels running between the sand-banks off those parts of the coast where the shore is flat and the tide ebbs to a considerable distance, I have repeatedly seen Swans apparently employed searching for food, marine weeds and grasses or small fish and crustacea being the only nutritious substances procurable. An old coast-gunner in the neighbourhood of Rye in Sussex assured me that he discovered a quantity of shrimps and other small fishes in a bird killed in the harbour; I did not, however, place much faith in the stories told by this character, many of his yarns concerning the birds he had seen or shot approaching the marvellous.

The thick coating of feathers on the body of a Swan and the size and strength of the wing-bone render these parts almost impenetrable to the charge from a 10- or a 12-bore shoulder-gun at the distance of five and thirty or forty yards. By bearing in mind, however, when a chance occurs, that the neck near the back of the head is the most vital spot, successful shots may occasionally be made at these distances. The use of heavy singles, especially 4-bores with excessively heavy charges, has lately been advocated by certain writers on sporting subjects; my own experience, however, is decidedly against these unwieldly weapons, and the following extracts from my notes will show that even Swans may be brought to bag by ordinary shoulder-guns. In the winters of 1858 and 1859 several of these birds were obtained in Pevensey Marsh and a few near Rye and Winchelsea; the whole, with but two exceptions, were killed with a 10-bore muzzle-loader, no larger shot than No. 3 being used. On the 19th of December 1858, while hidden behind a pile of old wreckage on the beach watching the endless swarms of fowl passing towards the west, a single bird flying along the coast was dropped perfectly dead by a charge of No. 6 shot at the distance of about thirty yards, the shot entering the upper part of the neck. On the following morning, while returning from flight-shooting at one of the sluices, a party of five old birds I had previously observed to fly in from the Channel again came in view, now making their way out across the level towards the coast. Drawing at once behind the shelter of the rough-hewn posts of an old marsh-gate, the approach of the Swans was anxiously awaited; though in the first instance holding a course that would have led them some distance to the east, they gradually edged round and passed within twenty yards of my place of concealment. The leading bird, which happened to be the largest, doubled up at once on receiving the charge of the first barrel (1½ oz. No. 3) in the body below the wing; the rest shearing off in confusion, the second was struck slightly too far aft, and skimmed on outspread pinions at least a couple of hundred yards before falling, when its wings were flapped for some minutes. Though no examination was made, it is probable that this bird was shot through the heart, its actions being almost precisely similar to those of one killed with a punt-gun a few winters back.

The distance that wildfowl will occasionally fly when fatally wounded is certainly astonishing; my notes for 1881 contain a reference to a bird that might reasonably have been expected to succumb at once to the injuries inflicted. On the 22nd of October a pair of old Swans accompanied by three young settled on a marsh adjoining one of the larger broads in the east of Norfolk; owing to a strong easterly gale it was no easy matter to work the gunning-punt within range, the birds having pitched directly to windward and an approach from any other quarter being impracticable. After watching the party for a couple of hours, during which they continued grazing on the marsh without shifting their quarters, I resolved, as the wind increased rather than diminished, to delay no longer in attempting to obtain a shot. To within the distance of about a couple of hundred yards a dense bed of reeds afforded protection from the gusts sweeping across the open marsh and intervening water; on emerging from this shelter the swell rolled over the deck and the force of the squalls rendered it almost impossible to make the slightest headway. After a lull of a few minutes we succeeded in working up to within a little over a hundred yards; and three of the birds happening to draw together, the heaviest cartridge that could be found in the punt-box (1-lb. B) was fired*. More by luck than skill, as the punt was rolling and pitching heavily, the aim proved true and the three birds were struck flat down on the marsh; a moment later, however, they recovered and, regaining their feet, flapped slowly off to windward. On closely watching their manner of flight it was evident that the shot had taken effect on one of the juveniles; with extended wings it turned and dropped to leeward, and after skimming about a couple of hundred yards rose a short distance in the air and fell headlong to the broad. Being forced to round a point of land, several minutes elapsed before we reached the spot, when the bird was still beating the water with its wings, its head being also raised and shaken; finally sinking beneath the surface as the punt drew up alongside. Having requested the taxidermist to whom the bird was sent for preservation to make an examination as to the cause of death, I was informed that three shots had passed through the heart.

A description of the colouring of the plumage of this juvenile is unnecessary, as the bird is figured in the Plate. After death I remarked that the beautiful pearl-grey tints on the feathers of the back and wings changed considerably, turning in some parts into a dirty grey or slate. This transformation probably accounts for the statement in one or two ornithological works, viz. "the young are brown in their plumage for the first year."

The wild and well-known call of the Wild Swan may be described as resembling the words "Whoop, Whoop;" the note, however, varies considerably in tone and strength, though wind and weather may have something to do with the sound. I never attempted to imitate their call, but have repeatedly seen these birds turn in answer to the shouts of the marshmen in the eastern counties; on one occasion half a dozen hovered round and finally settled on the water within a hundred yards of the bank on which our boat was brought up. Swans both by day and at night, when unsuspicious of danger, may be heard making a low chattering or grating noise, which would scarcely be audible at the distance of above forty or fifty yards. This sound might possibly be caused by the actions of the mandibles, though, to the best of my judgment, it appeared to resemble a low call-note.

* Swans were not expected so early in the season, or more suitable charges would have been in readiness.

BEWICK'S SWAN.

CYGNUS BEWICKI.

In his very interesting work 'A History of British Birds, &c.,' Seebohm gives (vol. iii. p. 484) the following lines referring to the discovery of this Swan:—"Bewick's Swan was discovered by Pallas nearly a century ago, but was regarded by the great Siberian traveller as only a small race of the Hooper. Naumann claims to have rediscovered it early in 1823 in Germany; Yarrell a year later in the south of England; and Hancock in January 1829 in the north of England."

The following extracts from my notes for 1871 refer to the shooting on Hickling Broad, in the east of Norfolk, of one of the few specimens of this species that I have succeeded in obtaining:—

"March 2. Over Heigham Sounds and the whole of the Broad early in the morning, and again towards evening. While in the centre of Hickling Broad, just after sunset, a wild Swan alighted on the water about one hundred yards from the punt, but would not allow our craft within range of the shoulder-gun, and the punt-gun was not on board. After three attempts, I refrained from disturbing the bird, trusting to fall in with it the following morning when better prepared.

"March 3. Down at the water at daybreak with the punt-gun, and met Nudd, the Hickling keeper, on the Sounds, out with his punt and gun. He stated that he had heard a Swan that morning and also the night before, but, owing to the thick fog hanging over the water, he had been unable to discover the bird. On reaching Hickling Broad, I made out a flock of fowl, which the natives had declared to be Tufted Ducks, on the deep water near the channel; but directly an attempt was made to scull towards them they rose, and after circling round pitched in the Smee corner. I then had the punt worked through the reed-bush; but immediately the open water was reached, at the distance of about one hundred yards, they rose on wing, when I ascertained the Tufted Ducks were Goldeneyes. I had noticed a Swan feeding with about fifty Coots while sculling to the divers, but, imagining it to be a tame bird, had not attempted to make an examination through the glasses. When the Goldeneyes rose, I jumped up to obtain a better view, and this movement started the Coots, and the Swan also getting on wing, I discovered my old acquaintance of the previous evening. The bird passed at about ninety yards, when I dropped down to the big gun and fired a charge of No. 1 shot; although evidently struck hard, the small stranger kept on across the open water, then on reaching the dyke towards the Sounds, turned and settled near the centre of the broad, with about a dozen tame Swans. After loading, we pulled well to windward and then sculled towards it in hopes of a shot. The bird, however, was exceedingly wary, and continued swimming before the wind till nearly on shore, when it rose again and flew across Rush Hill till almost out of sight, when it came round again and alighted in the open water about three hundred yards to windward of where our punt was stationed. As we happened to be in close proximity to the patch of rushes off Pleasure Hills, the punt was pushed among the reeds, and the man in charge of our second punt was sent off with orders to make an attempt to drive the bird in our direction. Just as he got round, the keeper, returning from the Sounds, came up in his punt, and the two boats had no trouble in forcing the bird towards where we were concealed. Before coming within range, nearly a dozen tame Swans which were being also driven before the boats came very close, and I fired just before the birds could join company. It was a long shot, but the bone of the right wing was broken close to the body, and it only required a short chase and a shot from the shoulder-gun to put an end to the sufferings of the unfortunate wounded bird. On examination, our prize proved to be a Bewick's Swan, a female in fine plumage, though only weighing 9 lbs. I was out again in the evening to make an attempt to obtain a shot at the Goldeneyes, to convince the natives that they were not Tufted Ducks, but did not happen to fall in with them."

The next entry in my notes is under the date of the 8th of December, of the same year, and refers again to these Swans on Hickling Broad. Commencing, however, with the 6th, the state of the weather will be better ascertained:—

"December 6. Not a single fowl, large or small, to be seen in the vicinity of the broad. We were only enabled with great difficulty to break through the ice to Hickling, to see the keeper, as the three nights' frost had almost made it strong enough to bear one's weight if walking across.

"December 8. To-day we commenced to cut a 'wake'* in the ice off Rush Hills, and, by help of saws and hatchets, cleared a space of about one hundred yards long and twenty wide. Not a fowl seen all day, with the exception of a single Swan, probably a Bewick's, which passed over the broad, flying towards the north-east.

"December 9. By mid-day we completed cutting the 'wake.' Very few fowl passed all day, only some half-dozen small parties of Divers that made their way in from the sea and, finding no accommodation, as the broad was all laid with ice, at once took their departure, and a herd of about fifty Bewick's Swans; these birds flew very low, but gave no signs of alighting, holding a straight course to the west. An unfortunate though somewhat amusing mishap occurred to-day while I had left for a few minutes a small thatched shed built up with willow-stakes and covered in with reeds, on the islet off the point of Rush Hills, to catch a glimpse of the Swans passing over the broad. There were ten or a dozen men at work and two, both farmers, who while out with their guns had stopped to assist us, had brought dogs with them. These sagacious animals had remained on the island while their masters were engaged at the 'wake.' It had been my occupation, assisted by one of the men, to attend to the preparation of our lunch, and a dozen tins of Cross and Blackwell's soup had just been heated in a large pan and made ready for use. On the Swans being reported in sight, I went out to ascertain, if possible, to which species they belonged, and to call in the men to take their lunch. Having satisfied myself that the birds were all Bewick's Swans, and evidently bound for other quarters, I returned to serve out the allowance, when we discovered that the two dogs left on the island had tilted over our pan of soup and consumed as much of the bones and meat as they could collect from the flooring of rushes on which it had fallen. Conscious of their iniquity, the delinquents had instantly made their way across the ice on the broad to Swim-coats, where they now sat on the bank surrounding the marsh watching us intently, and probably in expectation of a severe reprimand. One of my own retrievers, old 'Nell,' often referred to in these pages, had not moved from the position she had taken up near the stove when I left, and was evidently perfectly innocent of any participation in the crime, wagging her tail and exhibiting great delight at my return. Luckily a large stock of tins of preserved soups was at hand, and plenty of assistance being now procurable, a fresh supply was soon obtained, as only our soup had been interfered with. If I remember right, one of the culprits paid with his life for the offence, being shot by his master as soon as he came within range in a retired portion of the marshes on the way towards home."

My earliest experience with Bewick's Swan was in the winter of 1860, when I knocked down a fine bird of this species into the river running from the town of Rye, towards the harbour mouth. I had been out in a small rowing-boat, an old-fashioned tub, but luckily strongly built, with well-fastened timbers. My craft had

* A "wake" is the name given in this part of the country to a cutting made in the ice to obtain open water for the tame Swans.

been moored at slack tide to a strong post driven in the mud, when I mounted the river-bank to ascertain if any Snipe were about, and I was returning slowly when two Swans, evidently belonging to this species, were observed flying out from the inland marshes towards the coast. As soon as they came within range I was ready, and one dropped at once, falling about the middle of the river, and the other went off hard hit, giving signs of a fatal wound. On arriving at the river-bank I discovered that the tide was now sweeping down with considerable speed, and the bird was some hundred yards or so further down. Starting at once, I made the best of my way, but the blocks of ice tumbling over and rolling down were a great drawback, and the Swan, flapping and swimming with repeated strokes of its powerful paddles, was making good headway. A shot fired cut a lot of feathers from the head and neck; but the bird proceeded on with still greater speed, and the powder-flask* having been knocked from my hand by the force of a collision with an immense block of ice, fell into the water that the boat had taken in, and I was unable to fire another shot. The bird now rapidly increased the distance between us, and speedily flapped out of sight when the broken waves between the piers were reached. It was not an easy matter to reach the shore, but the difficulties were soon overcome; it is, however, quite possible that had it not been for the assistance rendered by the crew of one of the vessels lying at the wharf, my boat would have been swept out to sea.

While shooting a few Larks, one morning in January 1871, on the marshes near Shoreham harbour, in Sussex, a Bewick's Swan making its way inland from the Channel flew close over my head; having, however, only cartridges loaded with No. 10 shot, the charge had little or no effect. An hour before I had sent one of my boatmen into the village for some lunch and a supply of heavy cartridges; but unfortunately the latter were forgotten, and to this the bird owed its escape.

During the winter of 1868 I was staying at Tain, in the east of Ross-shire, for the punt-gunning on the Dornoch Firth; while returning home from the Meikle Ferry on a cold night in the large double punt close to the south shore, we had a narrow escape of what might have proved a serious accident. I was working the boat slowly along myself, and the puntman was lying forward half asleep, resting on the stock of the big gun which was trained ready for a shot. Just before we arrived off Morangie, and as soon as the darkness had commenced to set in, I detected a punt-gunner, conspicuous by his white jacket, rowing rapidly towards us. As our gun was pointing straight for his craft, I called to John the puntman, and ordered him to shift the barrel to one side. Without allowing me time to stop him, he put the hammer on full-cock and pulled the lanyard. Roused from his slumber, he imagined I had directed him to fire the gun at a Swan, in his half-stupified condition mistaking the white jacket of the puntman for one of these birds. Luckily the oakum with which we had covered the cap to exclude the damp had clung round the hammer, and there was no explosion, or it would probably have gone hard with the punt-gunner from the Meikle Ferry, at whom he aimed, and who remained perfectly ignorant of what had occurred.

* This was in the old times when only muzzle-loaders were in use.

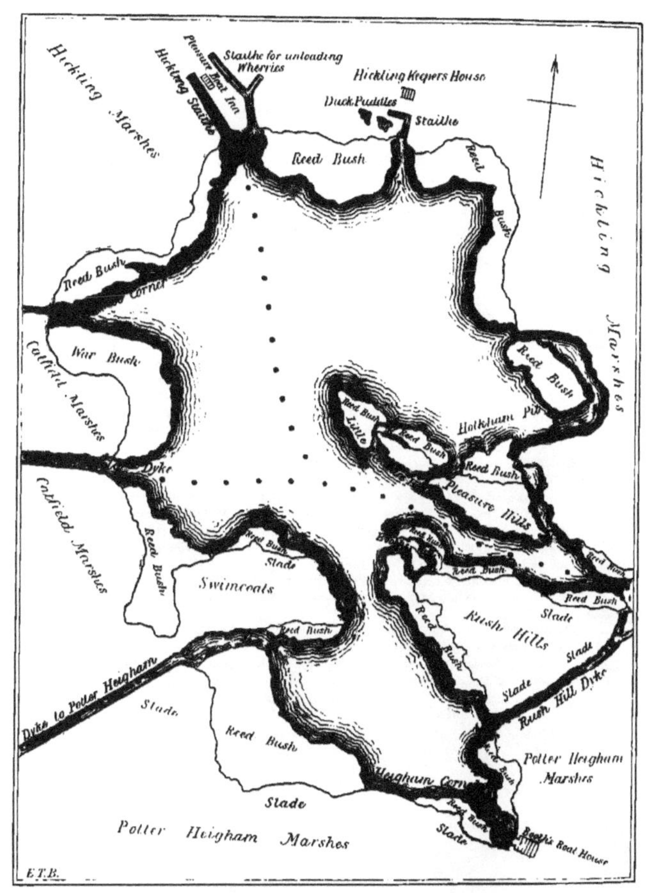

MAP OF HICKLING BROAD.

THE STAKES DRIVEN DOWN IN THE BROAD MARK THE COURSE TO BE FOLLOWED BY WHERRIES MAKING THEIR WAY TO HICKLING AND CATFIELD. IN OTHER PARTS THE WATER IS EXCEEDINGLY SHALLOW.

MUTE SWAN.

CYGNUS OLOR.

Though this species does not give utterance to loud cries like the Whooper, the name of Mute Swan appears scarcely applicable; let any one go afloat in a boat and drive before them the male and female with their brood, and they will learn whether the old birds cannot express their bad temper and indignation for the intruders by making a considerable noise. In Yarrell it is stated that "Swans were first brought into England from Cyprus by Richard I., who began his reign in 1189; and they are particularly mentioned in a MS. of the time of Edward I. (1272)." As this species has never been known to breed in a wild state in this country, it is probable that all our residents are the stock from the imported birds; possibly, however, as these Swans rear their young in Denmark and the south of Sweden, a straggler may occasionally visit our shores, as great numbers of perfect-winged birds are shot on the large pieces of water in several of the southern and eastern counties in severe weather.

As this species finds a place in every work on British birds I have examined, it is inserted in 'Rough Notes;' at different times I shot many full-winged birds that had never been pinioned or marked, which may have either escaped from some preserved water, owing to neglect, or made their way across the North Sea to our shores. Those I secured as specimens were a remarkably large and fine pair, that had spent their lives in peace and quietness, the sole occupants for many years of Somerton Broad, and were not known to have reared any young within the memory of those who gave me the information concerning them. They were shot on the 9th of November, 1871—the male weighing 32 lbs.* and the female 18 lbs., and both exhibiting most beautiful plumage. Among swanherds and those who are supposed to be learned in the management of this species the male is called a "Cob," the female a "Pen," and the young are generally known in all parts of the country as "Cygnets." I do not undertake to give all the changes through which this species passes during its course towards maturity; but I am of opinion that they take longer than the end of the second year, which is the time allotted to them in Yarrell, to reach the perfect adult dress and the full development of the colouring and protuberances on the beak. In the fourth edition of the last-mentioned work we are told, "The black tubercle at the base of the beak is called the berry, and a Swan without any mark on the beak is said to be clear billed."

The following extracts from my notes refer to Swans procured on the Norfolk Broads as well as to observations made in that part of the country concerning the behaviour of the unruly natives of the district:—

"November 29th, 1872. The keeper of Hickling who had charge of the Swans on the Broad and all the adjoining pieces of water had during the autumn put off the marking and pinioning of the Cygnets till they were strong enough to get on wing, and he was unable to capture them and perform the operation. The consequence was that the lawless gunners of the neighbourhood had gradually killed them down, till

* This may be considered a very heavy bird, 28 lbs. and 30 lbs. being given by one or two authors as the ordinary weight of the old male.

only two were now left on the water. These men had generally been accustomed to shoot them before daybreak, or at least on the quiet when no one was about. On this occasion, however, about midday they surrounded one young bird, and driving it towards the land, killed the helpless juvenile before it could rise on wing. The remaining young one was afterwards attacked in the same manner, but having been frequently fired at and ascertained their intentions he refused to allow himself to be driven, invariably flying off on the first signs of danger, and eventually escaping entirely from its persecutors and taking its departure for other quarters.

"June 21st, 1881. Received word of two Swans having been observed flying over Hickling Broad, and on proceeding to the spot I found them resting quietly on the water of Swimcoats, and had no difficulty in shooting both by a right and left with the shoulder-gun as they rose on wing at the approach of the punt. This pair of birds were doubtless immature, a few dark feathers on the plumage of their backs being detected here and there. The beaks were a pale flesh-tint, with a small black knob at the base and a black patch extending to the eye. Iris dark hazel. Legs light grey with dark veins showing. Webs deep grey, nails dark horn-colour. There was no telling whether these were what are termed Polish Swans, or only escaped tame birds that had evaded the operation of having their wings pinioned."

Again, under date of May 14th, 1883. "Two Swans circling over the open water on the Broad, and as soon as they settled we sculled up and secured them at once. They were probably in their second year. Beaks dark horn tint edged with black. Legs grey slate and the ridge edged with a deeper shade, almost black." The reason I always started in pursuit of the unpinioned Swans was because the keeper had requested me to kill them down, as there was then no excuse for the natives going and disturbing the water. I also imagined that strangers in various plumage might at times be obtained, but never clearly identified any worth recording.

On Hickling Broad, Heigham Sounds, and Horsey Mere I frequently shot birds that might pass for the form to which the name of Polish Swan has been given; but I am still of opinion that this supposed species is only one of the various stages through which the Mute Swan passes. The best description of all the forms and colourings of the beaks of these birds that could be written would give but little idea of the reality when the specimens are seen in life, and at least a dozen coloured plates would be needed to convey an impression of the heads and beaks as well as the plumage of the specimens I procured during the time I spent in this locality.

I often remarked on the Norfolk Broads that these birds are very pugnacious during the breeding-season, when each pair select their quarters and resent all intrusions of their neighbours on their haunts. A few years back I was at Potter Heigham in the spring, and had hired the shooting over the greater part of Hickling Broad. On going afloat early one morning I discovered a most terrible battle was being carried on between two fine old males; one of the combatants, I ascertained, had his quarters on Rush Hills, and the nest of his antagonist was built on the slade on Swimcoats. The commencement of the fight was not witnessed, and when I arrived on the scene they were holding one another down on the water, and striking violently with their wings whenever an opportunity occurred. Towards the end of their bout, and when both appeared fit to cry "Hold, enough"*, being completely worn out, the water was covered with feathers for a long distance, and had been in the same state all the time they were flogging one another with their wings. I do not think they exhibited their animosity for any length of time, as the following day the combatants were swimming peacefully round their respective quarters, without the slightest signs of an attempt to pick a quarrel again when either moved into the open water between the hills they frequented.

* The words "Hold, enough" occur in the inscription on the belt held by the champion of the ring, and the lines are taken from the last scene of Macbeth by Shakspere:—

"Lay on, Macduff;
And damn'd be him that first cries, Hold, enough."

COMMON SHELD-DUCK.

TADORNA CORNUTA.

MANY different names are bestowed on this species, and I have been rather puzzled to ascertain which was the best to give it in the pages of 'Rough Notes.' It is described under the heading of "The Common Sheld-Duck" in Yarrell, and this denomination appears the most appropriate, as by far the greater number of our wildfowl are termed Ducks, and not Drakes. It may not be out of place to mention the various titles by which Sheld-Ducks are referred to in a few of our latest published works on Ornithology, and to give the quaint appellations by which they are known to the natives in several remote parts of the British Islands. Yarrell, as I previously stated, prefers the name of Common Sheld-Duck, while Seebohm and Dresser call the bird the Common Sheldrake. It is also frequently spoken of as the Burrow-Duck by the natives of the locality in which it hatches its eggs. Seebohm, when referring to the various designations of Shield-drake, Shield-duck, Shell-duck, and Sheld-duck, states that "The name is derived from the low German *Scheldrak*, which may possibly refer to the shield-like protuberance at the base of the upper mandible of the bill; but Willughby and Ray stated, more than two hundred years ago, that they were called '*Sheldrakes* because they are particoloured.' In Norfolk it is provincially known as the Bargander, a corruption of Willughby and Ray's Bergander, a name borrowed by them from Aldrovandus, and obviously derived from the high German *Bergente*, though some writers interpret it as Burgander, 'bur' being a common north-country term for a burrow." The following extracts from Yarrell, describing the habits of this species and also stating the strange names they have acquired in certain remote localities, are doubtless correct, and the sources from which the information is obtained are given :—" Some are to be found on the sea-coast during the whole year, preferring flat shores, sandy bars, and links, where they breed in rabbit-burrows or other holes in the soft soil, whence the name of 'Burrow Duck' and 'Bargander.' In Scotland it is called 'Skeelting Goose,' according to Sibbald and other writers since his time, also 'Stock-Annet.' Many Sheld-Ducks come from the north to visit this country for the winter, for this species is rather intolerant of cold." Again, after making remarks on its breeding-habits along the eastern shores of both England and Scotland, the names it has acquired are referred to. To Shetland it is a rare visitor at any season; but is more common in Orkney, where, Dr. Patrick Neill says, "it has got the name of Sly Goose, from the arts which the natives find it employs to decoy them from the neighbourhood of its nest; it frequently feigns lameness, and waddles with one wing trailing on the ground, thus inducing a pursuit of itself, till, judging its young to be safe from discovery, it suddenly takes flight and leaves the outwitted Orcadian gaping with surprise." It is numerous in summer in the Hebrides, where it is known as the "Strand-Goose," and also in some districts on the west side of the mainland. Dresser does not give us much information concerning the nomenclature of this species, merely stating that the English names are "Sheldrake, Burrow-Sheldrake, Bargander." He, however, concludes his article by a short paragraph probably taken from the proof-sheets of the long-expected 3rd vol. of the 'Birds of Norfolk,' which he previously asserted had been placed at his disposal by the author :—" Mr. Stevenson, referring to the provincial name of 'Bargander' or 'Burgander,' by which this Duck is known in Norfolk, says that it is presumably a contraction of burrow-gander; but it appears to me just as probable that it is derived from its German name 'Bergente,' by which this bird is well known all along the German coasts."

In the summer of 1868 I was collecting in Ross-shire and Sutherland, and while staying at Dingwall, and constantly travelling by rail up and down the line that followed the course of the Dornoch Firth, I noticed that numbers of Sheld-Ducks frequented the waters of the salt-water firths and nested in the sand-banks extending along several parts of the coast. In June and July the young are following the old birds on the firths, and at my first attempt to secure specimens in this state I had a narrow escape from what might have proved a very unpleasant mishap. The following extract from my notes refers to the subject:—

"June 20. With three men I started by an early train from Lairg to Edderton, to go to the Meikle Ferry, to make an attempt to secure a brood of Sheld-Ducks that had been seen on the sands near the course we followed while passing up and down the line. On reaching the station and proceeding to the banks where a view could be obtained, we soon detected the whole family on the sandy flats near the shore and anticipated it would be easy to run them down. This, however, was soon discovered to be a mistaken idea, as the tiny mites were able to scamper and flutter over the ground nearly twice as fast as their pursuers, and I could only get near enough to stop one with a charge of shot, when the remainder of the brood had disappeared from sight. As the tide rose, the juveniles were observed out in Edderton Bay with the ducks, and after blowing out the india-rubber boat which had been brought with us I started in pursuit of them. The duck and a couple of young ones were soon shot, and I next made an attempt to follow the rest of the brood, when it was discovered that the blade of one of the oars had become detached and slipped off, leaving my craft to the mercy of the wind and tide *. Luckily the former proved most powerful, and my light and buoyant craft was soon forced over the ebb by the strong breeze, and I reached shallow water near Ardmore Point on the west side of Edderton Bay. Jumping overboard at the first chance, I dragged the boat ashore by the painter, and was none the worse for the mishap. The men were much put out on discovering that they were unable to procure any boats at the Meikle Ferry to come to my assistance; but they started off at once for the opposite side of the bay, a distance of about a couple of miles, when they ascertained that the wind springing up would carry my boat towards the shore in that quarter. Had the accident, however, happened an hour or two later, when the tide was running strongly out of the Firth, I might have been swept out to sea without attracting attention on either side of the water, unless the men had been able to give information at the fishing-stations." For the future I never ventured afloat in the india-rubber boat without stronger oars; these were made of ash, in one piece, and the remaining one of the old pair was fitted with a bolt run through the two parts and a nut screwed on the end, to act as a reserve in case of an accident.

On visiting the shores of Edderton Bay and making our way to the Meikle Ferry point a few days later, after a cold breeze had been blowing, we found three downy youngsters of this species lying on the shore at high-water mark, where the refuse cast up by the tide was collected. The helpless little mites were perfectly fresh and had probably only been washed up a few hours before; these doubtless belonged to the same brood as those that had perished from the effects of the cold after being deprived of the care and warmth afforded by the old duck that had been shot.

During a residence of three or four years in East Lothian, I discovered that Sheld-Ducks were very numerous about the Gullane Links in the breeding-season, these wide-spreading sandy flats sloping down to the shores of the Firth of Forth affording them ample accommodation for nesting-purposes. At times I also observed them on the waters of the Firth for many miles both east and west.

After a long continuation of severe weather in winter, birds of this species are frequently seen in the

* The oars were made in two parts, for convenience in packing, one fitting into a circle of tin on the other, and held together by a pin that was exceedingly liable to fall out.

Channel, off the south coast, in small parties of two or three or twice that number. In January 1881, after the disastrous gales that caused so much loss had swept over that part of the country, I shot an immature drake at sea, off Lancing in Sussex, on the 22nd of the month. The colouring of the soft parts was as follows:—Upper mandible a dirty light red tint, with pale shade of orange showing here and there, black ring round nostril, the nail dusky and a dark line round it. Lower mandible dirty livid flesh-tint. Legs, webs, and toes livid flesh-colour, very pale; nails pale horn-tint. On the 28th another was obtained off Shoreham; this was an immature duck, smaller, and the plumage not so brightly tinted. The colouring of the soft parts was almost precisely similar to that exhibited by the young drake.

To procure specimens of Sheld-Ducks in the most beautiful plumage, the adult birds must be obtained just before the breeding-season. After attending to the young and entering the burrows or holes in which their nests are placed, the feathers become worn and dirty by the constant rubbing against the sand or mould.

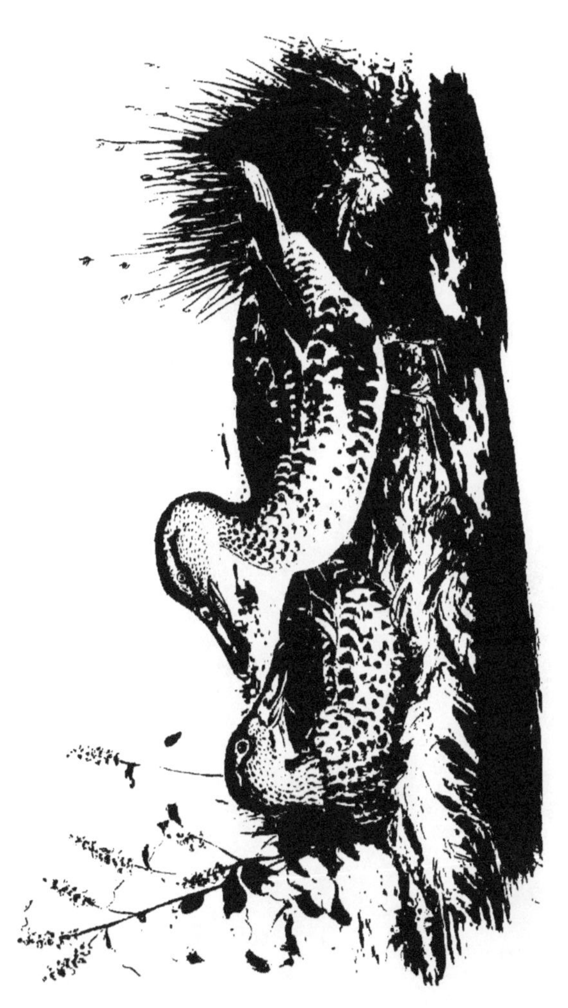

SHOVELLER.

ANAS CLYPEATA.

Though by no means abundant in any but strictly preserved districts, the Shoveller is widely distributed over the British Islands. In Sutherland I met with a few pairs and small flocks on the shallow pieces of water about Lairg, and on one occasion (June 1868) noticed a brightly plumaged male lying dead on the moors near Ben Armine. The rush-grown lochs in the east of Ross-shire are particularly attractive to this species, the character of the pools much resembling that of the Norfolk broads, where these birds are also resident. Further south than the swamps and flats of the eastern counties I have not detected their breeding-haunts, though Shovellers were annually seen during the winters I shot in Pevensey Level and Romney Marsh some twenty or five-and-twenty years ago. These birds appear to have decreased considerably of late years in the south of England, only two or three specimens having come under my observation since 1869.

Shovellers seldom gather in large flocks; from a dozen up to twice that number may, however, occasionally be seen on the Norfolk broads: if settling near other birds on water, or at the slades on the marshes, they appear to prefer their own society, and on rising again on wing almost invariably separate. This species shows little inclination for salt water: I never observed them resting at sea, after the manner of Mallard, Wigeon, Teal, or Pintail, off any part of the coast, though they occasionally alight among the marine weeds in muddy estuaries and feed round the shore, young birds being sure to occur on Breydon Flats in autumn. Here I have frequently met with them so early as the middle of August, exceedingly fearless, though, unfortunately, scarcely fit for table; to spare them, however, was utterly useless if a chance was offered, as many of the amateur gunners who frequent the water race immediately for any feathered creature that appears in view.

Some years back I kept three pairs of pinioned birds in confinement on a small island surrounded by wire netting on one of the Norfolk broads. Often when on the water at some distance I was surprised at the quick sight of the captives. On hearing their call-note, which is somewhat of a mixture between the cry of the Coot and that of the Moorhen, a small party of Shovellers might be detected high in the air; and though no answer was audible, the birds would invariably swoop down and, after circling for a time, alight near at hand. Many specimens, adult and immature, were obtained by the help of the decoys; fowl, doubtless of this species, were also often seen or heard, when passing the spot before daylight or after dark, to rise from their enclosure.

Shovellers are exceedingly tenacious of life, and when struck down wounded among weeds and water-plants much time is often spent before they are secured. A few lines extracted from my notes will give some idea of the extent of the injuries from which wildfowl will occasionally recover:—"April 25, 1873. A keeper who looked after the shooting over some of the adjoining marshes brought in a Shoveller which he had obtained the previous night, and believed to have escaped from the enclosure in which my decoy ducks were confined. The bird having exhibited signs of life had been carefully laid out among some moist rushes in a game-bag, and on removing the covering I at once recognized a drake that, having been imperfectly pinioned, occasionally was enabled to take a short flight from the enclosure *. To our surprise the bird commenced to move, and eventually stretched out its neck; though the skull being fractured, and part of the brain exposed, little hopes were entertained of its recovery. Having administered some water, the sufferer was placed in a corner of the duck-house, with all necessaries at hand; and fully expecting to find life extinct on our next visit, we left nature to take its course. Day after day, however, the bird appeared to regain strength, and after careful attention for some weeks it was enabled to stand and answer the call; in five or six weeks but slight signs of the injury could be detected. It was my intention to preserve the skull of this bird for examination, in order to ascertain the extent of the fracture; a few months later, however, it again took wing and never returned."

While in pursuit of specimens in March 1873 I was accidentally deprived of one of the finest adult males that ever came under my notice. At daybreak a pair were made out feeding among the leaves of the water-lilies on Heigham Sounds, and sculling quietly up within range, I fired, stopping both birds; on proceeding to the spot it was ascertained that the oakum in the punt-gun having struck the male on the neck, had completely cut away his head, which was discovered lying with the wad about twenty yards beyond the body. The felt wads now generally in use would render such mishaps almost impossible. Shot, when driven forcibly down by the rod, clings at times to the oakum (especially if the latter is damp), and the whole mass is carried like a ball †.

Drakes commence to lose their beauty towards the end of May, the feathers becoming worn and dingy; by the beginning of December I have seen several in full plumage, though many of the young males are scarcely to be detected from the females at this date. As to whether these juveniles would assume the mature dress before the end of the season, I cannot offer an opinion, having failed to rear these birds in confinement. On the 6th of December, 1872, a small party of Shovellers were sighted on Hickling Broad, and proving too wild to allow a near approach, a long shot was tried as they rose on wing. Among the slain were two or three young males, just showing a sufficient quantity of the red feathers on the underparts to proclaim the sex and distinguish them from some young females obtained the same day.

On the 1st of December, 1881, I fired the punt-gun at a dozen Shovellers on Hickling Broad, in the east of Norfolk, and stopped half the party; one drake had almost assumed the full adult dress, while another exhibited a most singular state of plumage. At first I imagined the bird to be a female, but on comparing it with one secured at the same shot the difference in colouring was at once obvious, the general tone of the plumage being far darker ‡. The colours of the soft parts were as follows:—Iris dull chrome-yellow (not quite so bright as in the adult male). Upper mandible olive-brown, with a patch of warm chrome-yellow at the lower part of the base; nail slightly darker; inside of nostril bright yellow. Lower mandible two shades—the upper portion olive-brown, the under orange. Legs and toes orange; webs dusky, with an orange line on each side of toes; joints rather darker; nails dusky horn. I am quite unable to account for such an unusual state of plumage being exhibited at this season; it is, however, possible that birds in this state may not unfrequently be procured and passed over as females.

The nest of this species is usually placed among patches of rushes on a dry part of the marsh,

* As a rule, I prefer a call-duck whose wing is clipped instead of pinioned: the latter is a cripple for life, while the former may be set at liberty when its services are no longer required.

† These remarks of course apply to the old muzzle-loaders.

‡ My opinion as to the sex of this bird was afterwards proved by dissection.

though at no great distance from a slade or swampy pool. In the summer of 1881 a Shoveller brought out her young from a stack of cut reeds piled up alongside a marsh-dyke in the east of Norfolk: the nest was placed about two feet from the ground.

A few words with reference to the Plates may be of service.

In Plate I. an immature male and female are shown in the plumage of the first winter; the specimens from which the sketches are taken were shot on Hickling Broad on December 6, 1872.

A male just assuming the full dress is represented in Plate II., together with the male in dark brown plumage previously referred to: both specimens were shot on Hickling Broad on December 1, 1881.

The perfect adult plumage of the male and female is given in Plate III.: the pair of birds from which the figures are taken were shot on Hickling Broad on December 5, 1872.

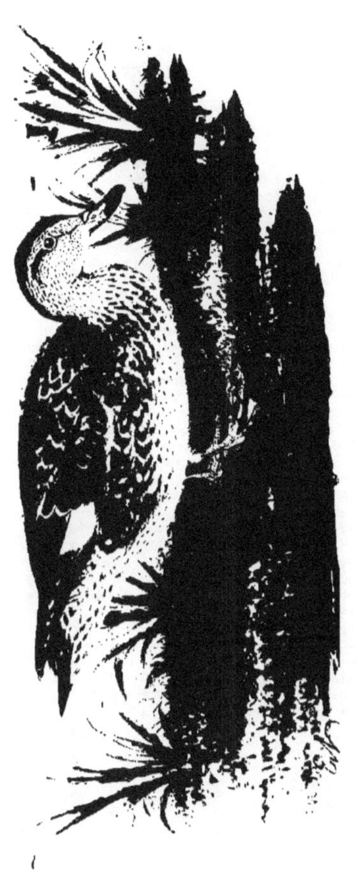

GADWALL.

ANAS STREPERA.

I have met with this species only on two occasions, both in the east of Norfolk. Some fourteen or fifteen years ago, in December, while Snipe-shooting on the old common at East Ruston, a couple of pairs were seen to alight on a broad water-dyke: a long and tedious circuit was needed to obtain a view sufficiently close to identify them with certainty, and then, owing to the want of cover and the waving bogs and swamps intervening, it was impossible to get within range for a shot. On the 15th of December, 1881, I noticed two fowl feeding busily on the fresh green grass round a small pool on Rush Hills, adjoining Hickling Broad, and a glimpse of them through the glasses as they moved among the tufts of rushes convinced me that they belonged to this species. As the distance at which they had settled was only about seventy yards out on the hill from the bank up to which the punt could be worked, I determined to try a shot with the big gun. Having picked out and decided on the nearest creek in the edge of the broad, we sculled in and awaited a chance for both to come together; there was but little delay, as they were continually passing from one spot to another while engaged in searching for food, and a shot was soon fired with good effect. One was killed on the spot, and the other appeared utterly helpless; before we could reach the spot, however, the latter rose on wing and fluttered slowly across the hill to the broad, where it immediately dropped on the water and paddled for a few yards, when it brought up and remained, occasionally shaking its head as if choking with blood in the throat. Not intending to lose a chance, a cartridge charged with small shot was inserted in the large breech-loading punt-gun, and sculling up within about fifty yards, the unfortunate cripple was at once obtained.

Both birds, as I anticipated, proved to be immature Gadwalls—young drakes in their first autumn plumage. As this species is far from common in the British Islands, and only the adults have been as yet figured by ornithological writers, one of these birds is represented in the Plate. There is no necessity to refer to the state of the plumage in these pages; but the colouring of the soft parts was as follows, the notes being taken down immediately after the specimens were lifted into the boat:—Iris dark hazel; a dark brown spreads halfway down the upper mandible in a line with the nostril on either side, a dark line also encircles the nostril; the lower portion and base of the upper mandible was a darkish yellow tint, and the lower mandible a paler yellow; nail a dusky white, and the ridge round the beak yellow; legs a dull yellow colour; webs dark grey, with a yellow line on each side of the toes; joints of toes and legs slightly shaded with a brown tinge, and nails a light brown tint.

Having met with so few chances of making observations on the habits of this species, I can only say that these birds appear to feed on the damp spots in the marshes and also in the water-dykes, where they swim about searching for food among the foliage overhanging the banks; whether they were engaged in making a meal off insects or vegetation I was too far off to ascertain.

Gadwalls breed in a few different parts of the eastern counties where they are well protected, and are often seen on some of the preserved waters in small parties of ten or a dozen.

MALLARD.

ANAS BOSCHAS.

Though seldom seen in such immense flocks as the Wigeon, the Mallard is perhaps the best known of all British Wildfowl. In every county throughout England and Scotland, and in most of the surrounding islands, this species breeds; and though their numbers have decreased of late years, there is still an abundant stock left.

I was doubtful for some time whether Mallard or Wild Duck was the most suitable title for this species; but as Yarrell, Seebohm, and several other authors use the former, it was adopted. Seebohm, in his 'History of British Birds,' when referring to the Mallard, relates many facts concerning the derivation of various names, and I have made a short extract from his work, bearing on this subject:—

"The formation of language is a process of evolution, and the meaning of words, when it has been definitely settled by custom, must be accepted without a too close inquiry into their derivation. The word horse is masculine, and mare is feminine; but when we speak of twenty horses it is not implied that there were no mares among them, the word horse being applied by custom in a special generic sense to include both sexes, but to exclude asses, cows, or any other quadruped. In precisely the same manner the word Mallard is used in a special specific sense to include both the male and female Wild Duck, to the exclusion of Shovellers, Pintails, or any other species of bird. *Malart* is a French word meaning drake, in contradistinction to *Canard*, which means duck. Possibly the word Mallard is a corruption of *mâle-canard*; but it has been used for a century to designate the species of Duck which is most common in a domesticated state, and may be applied to both male and female. In precisely the same way, when we speak of a Bean-Goose, we mean a definite species of Goose, irrespective of sex. It would be absurdly pedantic to speak of a flock composed of Bean-Geese and Bean-Ganders."

While residing at Ferrygate, a farm-house near North Berwick in East Lothian, to acquire a knowledge of agriculture, I soon ascertained that Wild Ducks were exceedingly abundant in that locality. The only means of getting a shot, however, was by awaiting their arrival at flight-time, the whole of the day being generally passed by the birds out on the open waters of the Firth, where they were utterly unapproachable. The potatoe-fields were one of their favourite feeding-grounds, and here several shots were often obtained, the hungry Ducks returning after being fired at, and others putting in an appearance till long after dark. On Gullane Links, a few miles further west, pits had been dug out in the sand round the pools to which the birds resorted, and sometimes very good sport was the result of a visit to this spot. In the hard winter of 1863, when passing the curling-pond at Dirleton, I saw the marks on the ice where a Mallard drake and two ducks had rested and been frozen on the surface. It was evident that the drake and one duck had been surprised and killed while held captive, as feathers and blood were on the ice; probably a fox or a dog had committed the murder. One cold night in December, the same winter, I was watching a frozen pool on Gullane Links at flight-time, close to a running stream, which was kept open and likely to attract fowl. There was a small gap in the bank, and having brought a rug, I sat down in the excavation to await the course of events. But few minutes had passed when a duck came in sight before I had time to fire a shot, and alighting on the ice glided right up to within a yard of where I had taken up my position. Jumping up suddenly, the bird was confused and, fluttering off, offered an easy shot, being conspicuous in the gloom from the light reflected by the snow and ice.

At some of the Highland lochs in the most unfrequented parts of the country there are many of these birds. On the 4th of July, 1868, while having our lunch on the banks of Loch Doula in Sutherland, my old black retriever "Nell" amused herself by exploring the dense reed-beds and slades near at hand, and captured and brought out sixteen full-sized birds, all in good condition. On the 26th March, 1869, a gale blowing from the east and the water rough and unfitted for shooting, we went out in the large punt on Loch Slyn, near Tain in the east of Ross-shire, and as we were returning after finding no birds, I happened to catch sight of the head of a Mallard in the heather on the bank of the loch. Though the punt was rolling in the swell, the gun was immediately turned on the bird, as I was anxious to get clear of the charge, which I imagined might possibly be damp from the spray that had broken over the boat *. As the trigger was pulled, the Mallard at which I aimed rose on wing, and at least a hundred others sprung up from the cover close at hand just beyond the bank. Luckily the one I first observed was in about the centre of the flock, and the charge passed through the thickest part of the dense mass. Many fell in the heather, rushes, and long grass, where they could not be followed; but we managed to secure about five-and-twenty, and as many more escaped in the thick cover, as my retriever had been left at a farm by the loch-side, as I had not expected she would be required. Before the punt had been stowed away a number of Grey Crows were observed, flying, screaming, and quarrelling near the spot where the shot had been fired. On returning it was ascertained that these destructive marauders had discovered and dragged out at least a dozen of the wounded birds that had managed to evade capture.

I seldom observed large flocks of Mallard on the Norfolk Broads in the east of the county, forty or fifty being sometimes seen, and on only one or two occasions as many as a hundred having been noticed in company. My observations in this part of the country only commenced in 1870, and probably the numbers of fowl had diminished as gunners had increased. In the autumn of 1873 a Mallard, knocked down while flying over the deep water on Breydon, dived among the green weeds and did not reappear on the surface; after waiting some time, and no signs of the bird being observed, the men commenced to stir up the weeds with their oars near where the bird had dived. After about half an hour's work he was discovered entangled in the strands of the weeds, which were exceedingly thick on that part of the water.

While at Innerwick, in Glenlyon in the west of Perthshire, we ascertained that numbers of Wild Ducks visited the marshes near the river during the night; there being little chance of obtaining a shot, traps were set in the pools of water the birds frequented. On reaching the spot the first morning it was evident that two birds had been taken, though only their beaks remained in the traps, the birds having been dragged away by some one who had observed them fluttering while attempting to escape. The farmer on whose ground this occurred was greatly put out, and wished me to request the minister to take the matter up and examine all those in the parish who were likely to be the delinquents, and make them confess their iniquity. I did not, however, take any steps to discover the culprits, but the traps which were set proved very successful, and they were never interfered with again.

I kept several pairs of Mallard in the enclosure with a pond in our garden at Brighton; no young were ever reared, and all in course of time died off after living in confinement for five or six years. I often watched one or two of the drakes in the summer running rapidly after the blue-bottle flies hovering round the pond and catching them with the greatest agility; they would also hunt out the black-beetles from the grass in moist parts of their ground, and gulp them down with evident relish, as though they were dainty morsels.

* This was in the old days of muzzle-loaders.

COMMON TEAL.

ANAS CRECCA.

The first of the Wildfowl family I ever shot was a handsome old drake Teal, killed when a school-boy from Harrow, home for the holidays, on a cold wintry morning at daybreak, just over thirty years ago. The bird was swimming in a small pool of water in a dyke that was partly frozen over in the marshes near the coast at Bulverhithe, between St. Leonard's-on-Sea and Bexhill in Sussex. This beautiful little fowl was considered a great prize, and the bag being presently augmented by some Ring-Dotterel and about half a score of Dunlins knocked out of a flock sweeping round one of the "pells"*, I returned in the evening considerably elated and determined to spend the rest of my vacation in pursuit of Wildfowl. After a time my expeditions were extended to Pevensey Level and inland marshes that stretch some miles up the country; these proved very attractive feeding-grounds for Mallard in those days, and great success was met with when I got more used to the work. A change, however, has come over this district, and now there is not one fowl to be seen where there were formerly hundreds of Ducks and Geese.

The Common Teal seems to be almost universally distributed over all suitable parts of England and Scotland that I have visited, breeding most abundantly in the heather on the slopes surrounding many of the remote Highland lochs in the wilds of Ross-shire and Sutherland. Large numbers also arrive on our shores in severe winters from the north of Europe and remain till driven further south by the long-continued inclemency of the weather.

While studying agriculture in East Lothian in the autumn of 1863 I had a narrow escape of a ducking, if nothing worse, through following a flock of this species. On my way from Ferrygate near North Berwick to Gullane, to dine with a farmer and afterwards await the flight of the Wild Ducks in the pits dug out at the pools on the links, I noticed half a dozen Teal drop on some small puddles of rain-water on the highest ridge of Ebris. This tiny rocky islet can be reached at low water from the mainland; and thinking there was time for a shot, I started off and succeeded in bagging a couple with each barrel. On reaching the foot of the rocks again, I discovered the tide commencing to flow over the low stony ground that had to be crossed before making the shore; and the water was up to my knees, but not over the long boots, before I arrived at the sandy links. On this very spot a man in charge of a cart and team of horses, who had been loading with stones on the shore, was caught by the tide a few years before, but how it happened none can tell. The carter was swept away by the force of the waves, and both horses were, I believe, drowned; possibly the cart had been overloaded, and the unfortunate animals were unable to drag it over the stony shore to the land.

While shooting on Breydon mudflats near Yarmouth, in the east of Norfolk, on the morning of the

* The pools of brackish water just inside the shingle-banks bordering the Channel are generally known by that name among the natives of this part of Sussex.

28th of August, 1871, the punts were drawn up under the wall on the north side, and soon after day-break a "bunch" of Teal came in view on the "deep water." Sculling up well within range, the trigger of the old muzzle-loader used in those days was pulled, but a dull click, which happened to put up the birds, was the sole response. On examining the nipple and cap they were found to be in perfect working order, and then the hammer came in for investigation, and the cause of the misfire was ascertained. In this the remains of four old exploded caps were discovered, and their presence would naturally take off the force of the fall and account for the loss of this excellent chance of a good shot. The breech-loaders that have since been turned out put an end to a chance of such mishaps, though had the lock of the gun been carefully attended to by the puntmen when they were cleaning it this could never have occurred. What had happened, however, proved of little consequence, as an hour later the small party of Teal were again detected floating quietly together off "Smith's Rondes," and five out of the six were stopped by the charge of the big gun, which was now in working order. Towards evening the remaining bird was secured, the poor little fowl having been evidently uneasy and restless during the whole of the day owing to the loss of its companions; just as we were on the point of leaving the water it flew in from the North Marshes and passing within range was killed by the shoulder-gun.

During the spring and summer of 1868 I collected a great many specimens in the west of Ross-shire, and while returning on the 22nd of May from a loch near the coast a few miles from Gairloch, where I had shot a pair of Black-throated Divers, we came across a duck Teal with her brood in a small reedy pool on a heather-clad slope on one of the hill-sides. As it was evident the juveniles would make good specimens of the young in the down, we halted under the cover of some dead reeds and slabs of stone and I extracted the cartridges in the barrels of my shoulder-gun in order to insert some loaded with No. 10 shot, so as not to damage such tender mites. One of these (a No. 10 bore pin-fire) dropped from my hand and striking a large stone the charge instantly went off, fortunately without appearing to have alarmed the birds or disturbed the equanimity of Nell, my retriever, under whose nose the explosion occurred as she sat at my feet and who paid not the slightest attention to the disturbance. The shot, when I fired, turned over the old Teal and the whole of her family, and Nell was despatched to retrieve them. This was satisfactorily accomplished, with the sole exception that having collected and crammed them all into her mouth at once, she delivered the old bird and seven downy young on the bank; but, despite all her efforts to bring it up, the last went down her throat, though she coughed and retched for nearly a minute.

I remarked, while living in the east of Norfolk, that rough weather in autumn and early winter was sure to bring us Teal on the broads and on the pools and slades on the hills surrounding the water. On the 1st of October, 1879, an exceedingly blusterous day, with squalls of wind and rain, there were a lot of Teal with Curlew, Plovers, and Dunlins on "Swimcoats," one of the hills on Hickling Broad, and some good shots were made, the birds appearing worn out by the protracted severity of the gale. As early as the 20th of June in 1883, I noticed many of the young birds of this species on wing on these waters, the old drakes still exhibiting the early spring plumage. Five males and two females were seen the same day on "Rush Hills"; this would indicate that they had either lost their broods or separated from the young, which were now capable of attending to their own wants, the latter being most probably the case.

I have kept Teal in confinement in our garden near Brighton for several years, but have been enabled to learn little concerning their habits, as they appear exceedingly shy, usually seeking the shelter of the willows and other bushes as soon as they detect their actions are watched. Under the date of April 11, 1881, I find in my notes referring to these birds, "The Teal drakes whistling almost continually when not disturbed, the sound being somewhat similar to that of the call of the cock Bullfinch. The note of the female, who now and then responds, resembles the squeak of a penny trumpet." Also on the 18th of

July the same year the following is inserted:—"Females very noisy, indulging in their penny-trumpet performances."

The following are a few extracts describing the changes of the plumage of the captive drakes:—

"July 14th, 1884. Teal drakes showing first signs of moulting."

"July 16th, 1884. Flanks of Teal drakes considerably changed."

"July 30th, 1884. Plumage of drakes very much speckled, almost lost all signs of male attire. The red on the feathers of the head and the yellow lines had totally disappeared from the brightest old male, and the grey markings on the back had also undergone a great change."

"August 31, 1884. Smaller and darker markings on the plumage of the drake in summer than on that of the female; the pale cinnamon patch near the tail, however, also indicates the sex at this season."

"October 18th, 1884. The outline of the red colouring on the head of one of the drakes just defined with the markings, but very imperfect; the rest of the plumage clouded as in summer."

WIGEON.

ANAS PENELOPE.

There is no denying the fact that in days gone by Wigeon visited our shores in larger numbers than any other of the Duck family, and doubtless they still return where allowed to remain unmolested. The increasing numbers of punt-gunners and shore-shooters have, however, at the present time left but few undisturbed quarters in which the large flocks of this species may now obtain a plentiful supply of food and quiet repose.

Wigeon breed in several of the northern counties of the Highlands, and are so abundant in some parts that, on one occasion, I remember fifteen nests were passed and examined in a collecting expedition that occupied us for two or three days in a remarkably wild and remote district that is never likely to be intruded on by those who would interfere with them. In June 1868, we found several broods of young on the flat moors stretching towards Ben Armine and other high hills in Sutherland, and after a long tramp we had our lunch at a shepherd's shealing. Among other dainties supplied, in addition to those we carried, the good wife helped me plentifully to an immense piece of sheep's milk cheese; this curious concoction might possibly have been very palatable, had it not been spoiled by being crammed with carraway-seeds, which gave a most peculiar and unpleasant flavour. Fortunately, with the assistance of a collie dog lying at my feet under the table, who snapped up and swallowed what I dropped to him, the whole was disposed of, as my refusal to consume this delicacy might possibly have given offence.

This species, like other wild fowl, occasionally suffers from floods in the Highlands; a short extract from my notes for 1868, while collecting in Ross-shire and Sutherland refers to the subject:—

"June 6, 1868. A tremendous storm of rain followed by floods came on during the night, and must have destroyed the eggs of thousands of Gulls, Ducks, and Divers in the country around Loch Maree. The water in most of the lochs rose several feet, and the greatest number of nests are placed on the banks or out on the open moors only just above the ordinary surface of the water. About an hour before mid-day, as the violence of the storm gradually diminished, and it eventually became merely a drifty Highland mist, we drove part of the way, and then ascended to Loch Clare to take the nest of a Wigeon the keeper had seen the day before. Not a Wigeon was to be seen on Loch Clare or Coulan, and the nest was discovered to be about two feet below the surface of the water, the loch having risen four feet at least during the night. The nest and eight eggs were, however, fished up, and we then made tracks back over the moors and returned to the inn at Kenlochewe, as it was no use searching for the birds. On the previous day there had been several Wigeon ducks and about a dozen drakes on the loch, all of which most probably had their nests or broods near at hand; now they had all taken their departure. The floods had carried away a bridge, and we had some difficulty in crossing one or two of the burns, all of which were greatly swollen. We did not reach Gairloch (which was our headquarters) till late at night, having been away two days and never fired a shot."

There are many notes referring to Wigeon in my journals, when staying at Tain in the east of Ross-shire, that describe the numbers met with on the Dornoch Firth. How confiding and unsuspicious of danger this species is at the commencement of the shooting-season, and when first arriving on our shores, would scarcely be credited except by those who have witnessed their actions.

"October 10, 1868. Immense flocks of Wigeon on the firth and also on the flats at low water. The birds proved exceedingly tame, running and paddling round the punt when we took the ground, and were left on the sands by the ebb tide before moving to get our craft afloat again, as we had remained quiet in order to make observations."

"December 7, 1868. Great numbers of Wigeon further up the firth near Invershin, the big pool some distance above the railway bridge being almost covered by a mass of birds."

"March 4, 1869. Between three and four hundred Wigeon on the Kyle * nearly opposite Inveran, some on the large blocks of ice, many swimming on the water, and several cleaning their feathers on the banks, at the side of the river. As we came up in the punt to procure some of the Wigeon, the ice broke into several parts, and at once drifting down with the tide, divided the birds and spoiled our chance of making a heavy shot. We kept on till a large block of ice, rolling round, was just about to strike the bows of the punt, and then I was forced to pull the trigger and only six-and-twenty were collected; had we arrived on the spot a few minutes earlier, twice that number might easily have been obtained."

On the 11th of March, 1869, I was at the Little Ferry, a small muddy harbour near Golspie in Sutherland, and while coming down the channel on the ebb from the upper part, we sighted some Wigeon spread out on a mussel-scarp, and all resting quietly. Having noticed them in time, I was enabled to make preparations, and when we arrived within range the punt-gun was discharged, and only one from the outside of the line got on wing, and he was speedily turned over by the 10-bore breechloader, and a cripple that attempted to make off by diving into the water was stopped by the second barrel. When the slain were collected, they proved to be fifteen Wigeon, all drakes.

I have occasionally known Wigeon to remain on Hickling Broad all through summer; these birds always exhibited immature plumage, and were doubtless either weakly or had suffered from wounds, and were unable to follow their stronger relatives. The marshmen were of opinion that they were Garganey Teal, till I shot one or two, in order that their identity should be established.

On a few of the Norfolk Broads in the east of the county these birds are known among the natives of that remote district by the name of the "Smee."

Bewick, in his 'History of British Birds' (published in 1804), gives a short, but quaint and amusing description of the habits of this species :—

"The Wigeons commonly fly, in small flocks, during the night, and may be known from others by their whistling note when they are on wing. They are easily domesticated in places where there is plenty of water, and are much admired for their beauty, sprightly look, and busy, frolicsome manners."

I have kept several Wigeon in confinement, but as there were never any signs of their nesting, I tried other species. Nearly all were as tame as the Wild Ducks, though a few proved timid and sought the shelter of the bushes and shrubs round the pond whenever the enclosure in which they were confined was visited. They were exceedingly fond of snapping up any insects that came in their way. I noticed that a duck, standing on the edge of the pond in company with a drake, on the evening of the 30th of June, 1884, turned her head and ran rapidly to the rough grass under a clump of willows near which I was sitting watching the wildfowl, and caught a fine poplar hawk-moth, which was crawling up the stems of the grass, and devoured it. I see by the entries in my notes that these birds also captured and swallowed the daddy-long-legs hovering about near the pond, during September the same year.

* The "Kyle of Sutherland" is described on Black's map of Scotland as extending from some few miles above Inveran to about as many miles below Invershin.

Here is a note referring to the well-known call of the drake Wigeon :—"February 22, 1884. Drakes commencing to whistle. The mandibles are opened wide for several seconds before a sound is emitted." It is also stated that these birds were particularly noisy about the end of October, calling continually for several hours when their haunts were not intruded on by strangers; as I was a frequent visitor, little notice was taken of my presence.

LONG-TAILED DUCK.

ANAS GLACIALIS.

THE Long-tailed Duck has been met with in larger or smaller numbers on all the seas, firths, and saltwater lochs off every portion of our coast-line where I have spent any time in studying the habits of Wildfowl and Sea-birds. This species also occasionally penetrates to inland waters; early in 1872 I examined an immature bird shot by a man who worked in the marshes on Hickling Broad in the east of Norfolk.

The various species of fowl that procure their food by diving are seldom seen on land unless during the breeding-season. I have only on one occasion watched Long-tailed Ducks exhibiting their walking capabilities. While lying off the southern shores of the Dornoch Firth, halfway between Tain and the bar, one afternoon in the winter of 1868, five fowl, which I failed to recognize, were observed moving about and apparently feeding round the pools on the sands at the distance of between three and four hundred yards. A glance through a strong binocular almost convinced the puntman and myself that the unknown were Long-tails; so, directing our craft to be put ashore, I made my way inland across the wide stretch of sands, holding a course that would lead past the birds at about ninety or one hundred yards. On drawing nearer it became evident that our supposition was correct, the small group consisting of a couple of drakes and three female Long-tails. It was not till I advanced within an easy gunshot that the birds rose on wing, and flying straight for the open waters of the firth, alighted about half a mile from the shore, and dropped slowly up the channel with the flood-tide. The whole party proved as confiding on the water as on land, and allowed us to scull past within thirty yards, affording excellent chances for observation. Though the drakes had each the two long tail-feathers well developed, their plumage was scarcely perfect, and not needing specimens in that condition we left them undisturbed. I remarked that the tails, as depicted in Gould's work, were carried at a high elevation, occasionally drooped, but for the most part lifted up in a most jaunty manner. As the daylight faded while we waited for the tide to flow up the course of the small river running through the sands towards our boat-house on the waste ground below Tain, their calls were plainly audible; the sounds arising from various quarters, the birds were evidently being answered by others further out in the firth. The note of the Long-tailed Duck is most peculiar, one of the local names of this species, "Coal and Candle-light," being derived from a resemblance it is supposed to have to those words, which the bird pronounces in a sing-song manner. The combined cries of a flock of these birds resounding over the scarcely audible ripple of the waves is exceedingly impressive when heard during the hours of darkness on fine still nights, and is sure to attract attention when listened to for the first few times.

While punt-gunning at the Little Ferry near Golspie in Sutherland early in March 1869, I found the Long-tails very numerous, flocks of from ten or a dozen up to a score often being observed fishing in the channel running into the harbour, and a few occasionally making their way to the saltwater loch at high water; and here I procured some drakes in fine plumage. These birds fly with great speed, and I had excellent chances for judging the pace when watching them from the shingle-banks as they dashed up and down over the narrow channel leading to the harbour. How many feet it was necessary to hold ahead to ensure a successful shot it is hard to say, as the exact distance to be allowed in order to stop a bird at full spin was not arrived at till after repeated failures. I found all the Long-tail drakes in full winter plumage when visiting this interesting locality, in the second week in March 1869, for a few days' shooting, and after returning a month later I noticed the old drakes had undergone a considerable change. Many had assumed about half perfect plumage, while others were far advanced, and a few in quite full summer dress. In these stages they seldom made their way over the bar into the channel towards the harbour, but kept, as a rule, to the open firth at the distance of about a quarter of a mile from the shore, drifting down occasionally as far as Golspie. Velvet Scoters were also plentiful off this part of the coast, and the two species often intermixed while working up and down the firth: when alarmed, however, and put on wing they invariably separated, each following a course of its own. In order to make observations on the various stages of plumage exhibited by the drake Long-tails, I left the harbour in a large double gunning-punt and proceeded down the firth a mile towards the east, and our cruise having come to a successful termination, we were amply repaid for the risky start we were forced to make. The current over the bar ran strong, and the seas caused by the tide were somewhat lumpy; but we reached the open firth without shipping a drop of water or receiving a sprinkle from the spray. Cork jackets and life-belts with a line and buoy for the punt-gun had, however, been provided in case of a mishap. Adult drakes almost if not quite perfect, and some in less advanced plumage, were obtained, with one or two exceedingly handsome males of the Velvet Scoter. Coloured sketches also of the heads, beaks, and legs and feet of each were taken before we turned back, in order to reach the harbour with the flood. Having been followed by a fishing-craft with a strong crew, we were amply provided for in case of bad weather setting in or squalls blowing up unexpectedly from the wild North Sea.

Immature birds in considerable numbers as well as a few adults annually work their way south on the approach of winter, and take up their quarters off our eastern and southern coasts, for the most part selecting such situations over mussel-banks and other feeding-grounds that are suitable to Scoters and other diving ducks. On the 28th of November, 1879, while steaming along the shore and also through the gatways between Yarmouth, Lowestoft, and on towards the south as far as Benacre and Southwold, I met with several large flocks near each of the places mentioned. Common Scoters were by far the most numerous of the fowl gathered off this attractive portion of the coast, though Velvet Scoters, Long-tails, and Goldeneyes were well represented, and one small party of three or four Scaup were sighted in the Ham to the south of Gorlestone pier. The sprat-boats were making enormous hauls near the Pakefield gat as well as off Kessingland and Covehithe; thousands of Kittiwakes with a few immature Pomatorhine Skuas in attendance were hovering round the nets; but I doubt if the fish were any attraction to the ducks, as they were only observed diving for food over the banks. Every winter, when spending any time at sea off that part of the coast, I have observed Long-tailed Ducks; it is, however, only when the weather is severe that they favour the flat sandy shores to the west of Brighton in Sussex with a visit. In December 1879 and again in 1880 I remarked them in greater numbers than usual, and in the former year met with repeated opportunities for examining a handsome male who appeared in perfect plumage, with the exception that the two long feathers of his tail were absent. I noticed him first on the 7th of December, and for a couple of months after that date he remained along the coast between Shoreham and Worthing, defying all attempts made to secure him by shooters in open boats, though paying little or no attention to our double punt passing while he was engaged in diving over the banks. Small parties of from six or eight up to double that number are not unfrequently seen in this part. On a fine still morning in the last week in December 1883, when the sea was as smooth as glass, I noticed about a dozen in company with as many Eiders and some hundreds of Common and Velvet Scoters; so watchful, however, were the whole community that it was utterly impossible to approach within range.

PINTAIL DUCK.

DAFILA ACUTA.

This elegant species appears to be distributed over all suitable parts of the British Islands I have visited. While punt-gunning on the firths intersecting the north-eastern shores of the Highlands, in the winters of 1868 and 1869, I met with considerable numbers, most being seen in early spring. Small parties also occasionally put in an appearance on the Norfolk Broads, and a few resorted to the rivers or flooded portions of the flats of Pevensey Level and Romney Marsh during the winters I passed in those districts.

I never succeeded in discovering the nest of the Pintail in this country, though I find in my notes for 1878, under date of May 28, that a pair were often observed on and around the hills on Hickling Broad, in the east of Norfolk, and that latterly the drake (an exceedingly brightly marked bird) was usually seen alone. I also ascertained that, a short time previously, a duck's nest with eight eggs, supposed to be those of a Shoveller, had been taken by a marshman on the same hill these birds frequented. As it is improbable that the natives were well acquainted with the eggs of either species, this clutch might possibly have belonged to the Pintail.

While observing the habits of the fowl on the Dornoch and Cromarty firths, I find by the entries in my notes that the first arrivals of Pintails usually took place about the middle of October, and a few birds were often obtained about that date while flight-shooting on the Tain Sands. It was not till the close of winter or early spring that flocks of this species were observed on the waters of the firths. I never noticed large mixed bodies of males and females, seldom more than ten or a dozen being in company when both sexes were represented, though thirty, forty, or even fifty drakes were often met with by themselves. On one occasion (March 15th, 1869), in the Dornoch Firth, I had sculled a small single punt within about one hundred yards of at least a score of fine drakes, and was on the point of firing the big gun, in order to obtain some specimens, when the whole were unexpectedly put on wing by several Great Black-backed Gulls swooping over with loud screams. These rapacious Gulls having previously been accustomed to secure the cripples disabled by the gunners, were by no means satisfied by their first attempt to secure prey, and continued so constantly in attendance that all sport for the day was at an end, till I turned my attention to thinning down their numbers, when the few survivors gave the punt a wide berth.

On this part of the coast the few Pintails that remain throughout the winter often join with the immense flocks of Wigeon, and are exceedingly conspicuous when the dense mass of fowl are feeding over the weed-grown portions of the sandy flats, their superior size and long white necks at once attracting attention. The local punt-gunners have bestowed the name of "Wigeon-leaders" upon these birds, on account of their generally being a little in advance of the front ranks of the flocks; when on wing also they are invariably in the van. On the 7th of April, 1869, I fired a shot with the big gun at two pairs of this species, floating with the tide up the Dornoch Firth, off Morangie, and having secured the four birds, was returning to the shore to load, when several seals were noticed within a few hundred yards. Having recharged, I was speedily in pursuit of two or three that appeared inquisitive, and a successful shot having been made, the carcass of the unsuspecting denizen of our firths was dragged to the nearest sand-bank, and at length, after considerable labour, stowed away under the stern-deck of the punt. On my way back, another duck and drake Pintail came in view, and while sculling up for a shot, the sound of cracking timbers caught my ear, and on looking round I discovered that the supposed defunct seal showed signs of life, and having raised his head and shoulders, the after-deck was much strained and almost wrenched from its fastenings. Well aware that no time could be lost, I picked up a small breech-loading rifle, and a bullet through the head at once restored quiet; luckily the water was not above two feet in depth, and the sands were soon reached in safety without further mishap. This seal proved to be exceedingly heavy, and yielded, when the blubber was cut off and boiled down, just over three gallons of oil *.

A party of Pintail drakes when feeding on a rush-grown pool, with their heads straight down under water and their long pointed tails elevated in the air, present a most singular appearance. I have frequently watched numbers engaged in this manner on a small muddy loch near Fearn, in Ross-shire, that was almost over-stocked with eels †.

Though usually consorting with Wigeon, this species at times joins in company with Mallard. I find in my notes under date of November 4, 1871, while shooting on Hickling Broad, in the east of Norfolk, that, after having placed out several of our floating dummies ‡, a small party of about a dozen fowl wheeled round and pitched near at hand, allowing the punt to scull up within range before they rose on wing; on collecting the five that fell dead to the shot, they proved to be four Wild Ducks and a Pintail Duck. Though usually watchful, I have on two or three occasions remarked that this species appeared entirely off their guard. While sculling up to a mixed flock of Wigeon and Pochard, on Hickling Broad, on the 5th of March, 1873, I passed within ten yards of a couple of pairs of Pintails—ducks and drakes both keeping their beaks covered in the plumage of their backs, and making not the slightest attempt to rise, or even to paddle further out of our course.

* A reference to the seals shot in this firth, and the manner in which the oil extracted was disposed of, will be found on page 3 of the Brent Goose.

† In this district, as well as in many other parts of the Highlands, the natives evince a strong and incomprehensible dislike to an eel: consequently this excellent fish remains free from persecution, and increases in size and numbers. At this small piece of water (known in Gaelic by some unpronounceable name signifying the "muddy loch") the eels, at the time of my visit in the spring of 1869, were positively swarming in the shallows; and while wading through the soft mud and reeds, from one bank or point to another, several, apparently from five or six pounds in weight, or possibly even considerably heavier, would be driven from the shelter into which they had made their way. After causing a great commotion beneath the surface and among the weeds, they would eventually disappear in the depths of these almost bottomless swamps.

‡ These were wooden models representing Ducks, carved out of yellow pine, and well coated with paint in order to exclude the damp. It was somewhat singular that the fowl attracted by these dummies did not take the slightest notice of their colouring. The men who constructed them were by no means artistic, and I had intended to give a few finishing touches before they were made use of; it was, however, soon discovered that this addition was perfectly unnecessary. In order to keep our decoys upright when afloat, a leaden weight was fixed under the belly, and a ring screwed into the breast to hold the anchor-line for mooring and to enable the wooden birds to ride head to wind. The deception is then perfect, and many a charge of shot have our dummies received from passing gunners when left unguarded on waters where the shooting was not preserved.

GARGANEY.

QUERQUEDULA CIRCIA.

Though his colouring is unpretending, there are few, if any, of our Wildfowl more beautifully marked than the drake Garganey. This species appears to have been seen and also obtained in several English counties, and on a few occasions in Scotland and the outlying islands; I have, however, only met with it in the Broad district in Norfolk. These handsome little birds are still numerous in that part of the country, arriving early in the spring, taking up their quarters and rearing their young in the neighbourhood of the largest piece of water, and making a move to a more suitable climate before cold weather sets in. On two or three occasions I heard of Garganeys shot in winter, but in every instance where the strangers underwent examination by competent judges they proved to be female Wigeon.

My own experience with regard to the situation chosen by this species for its nest differs considerably from the statement in the last edition of Yarrell, that "in the Broad district in Norfolk, the densest reed-beds are preferred." About Hickling Broad, where I have had ample opportunities of observing them during the summer, I remarked that the eggs were usually laid in the patches of rushes in the unreclaimed marshes, at some little distance from the water, not a single nest having, to the best of my knowledge, ever been detected in a reed-bed. Now and then the birds were known to have bred among the long coarse grass and tufts of rushes on the dryer portion of the hills surrounding the broads, but, as a rule, they go further from their usual haunts.

While staying at Potter Heigham, in the east of Norfolk, in the summer of 1883, I was sent for early on the morning of the 16th of June, by one of the natives of the village, who had been on his way to market and surprised a brood of wild Ducks in a ditch by the roadside. The weather at the time was rough and stormy, with rain falling in blinding torrents, and on reaching the spot, which luckily was only about a quarter of a mile from the farmhouse at which I was staying, a female Garganey was seen, standing with her neck stretched out, in the middle of the road. Two or three times she ran towards the ditch, which was deep and overgrown with brambles and wild plants, and then withdrew slowly to her former station. On searching the cover, eight young ducklings were soon found, and placed in a basket wrapped up in flannel, the poor little mites being wet and weakened by exposure to the rain and the damp grass through which they had been led. As the old bird, after flying round in circles over the adjoining fields, still continued to return to the road when we had left the spot, to take the young to the house in order that they might have better attention, I returned again and took up a position to watch her movements. Her actions soon led us to believe that some of the brood were still at large, and on again turning over the ferns and rubbish in the ditch, two more downy youngsters were secured, a hundred yards or so nearer to the farm than those previously obtained. It was not, however, for two or three hours that the poor old duck deserted the spot, returning again and again, after flying round, and alighting either in the fields or on the road near the ditch; she appeared much distressed by the loss and perfectly regardless of danger on her own account, seldom attempting to rise on wing till approached within four or five paces. The road on which these young ducks were met with was at least a mile from any rush-marsh or other suitable locality where a Garganey might have been supposed to nest, and was bordered on either side by cultivated fields. A large pollard oak was overhanging the spot where the birds were first observed; and as it has been stated that Wildfowl have occasionally been known to breed in the shelter to be obtained where the branches have been cut, I carefully examined the whole of the crown of the tree, but not the slightest signs of the young having been there could be detected. It is probable that the old bird was taking her newly hatched brood down towards the Broad, but where her nest could have been, unless in some dry hedgerow, it is impossible to form an opinion. When first brought in the young birds were extremely lively, and on being placed in a box, about eighteen or twenty inches in height, lined with flannel, two or three made their escape, climbing up the side with the greatest rapidity, by aid of their sharp little claws, and commencing to catch flies on the panes of glass in the window. A few hours after they had been taken the youngsters appeared to grow gradually weaker and become more helpless, finally gasping for breath, and though the greatest care and attention was bestowed on them, none survived the day.

There is little difference in the colouring of the down on the nestlings of the Garganey and Common Teal, both exhibiting much the same yellow tinge, with dark brown markings. The colours of the soft parts of the young of this species may be described as follows, the notes being taken while the birds were alive:—Upper mandible dark brown, nail red-brown, under mandible flesh tint; inside of mouth pale flesh; iris hazel; legs dark brown, a light line outside leg and at the side of each toe, nails a dull brown.

In 1875 I noticed several broods of Garganey were strong on wing by the end of the first week in July, numbers coming every day to feed about the pools in the reed-bushes at Hickling Broad and affording capital sport if a few were required. In the state of plumage exhibited at this season immature Garganeys have more than once been mistaken for the Blue-winged Teal, and, considering that the juveniles have never been figured by any author, the error is certainly excusable in those who have never met with a chance of observing them in life. The colours of the soft parts of those shot at this time were as follows:—Upper and lower mandibles a dull smoky blue; legs, toes, and webs of feet the same tint; iris hazel.

EIDER.

SOMATERIA MOLLISSIMA.

Though no records exist as to the breeding-range of the Eider having at any time extended far south of the Fern Islands, off the coast of Northumberland, scattered parties of these birds may occasionally be met with in autumn and winter off all parts of our southern and eastern shores. At this season they are shy, and, usually rising at long distances, the identity of the flock is seldom ascertained. I refer to this fact as Eiders have repeatedly come under my notice off the Norfolk, Suffolk, and Sussex coasts, though but few instances of the occurrence of the species in these quarters have been reported.

The numerous breeding-haunts of the Eider round our northern coasts have been so often referred to that it is superfluous to give a list of those I visited. Though the stations on the islands in the Firth of Forth are enumerated, the sandy links of Gullane appear to have escaped the notice of writers. I discovered but one nest on that lonely stretch of ground, though at daybreak and late in the evening the drakes might be watched flying over the sand-hills; and this fact was considered a sure sign by the natives (all noted egg-stealers) that the females were sitting near at hand. On making inquiries when staying at Canty Bay a few years back, I learned that the links had been entirely deserted and that but few young birds were now reared on the rocky islands off the coast. The Bass is mentioned by certain authors as a breeding-station; the constant stream of sightseers has, however, driven the Eider from this part of the Firth, and on only one occasion during the last twenty years have eggs been laid on the rock, the site then chosen being among the ruins below the fortifications on the south face. The ledge to which the bird resorted (encumbered with large blocks of fallen masonry, and luxuriant in summer with the attractive foliage of the sea-beet and Bass mallow *) was at a considerable height above the water, with a sheer descent of not less than fifty feet; under such circumstances it was evident that the young would need assistance from their parents to reach their natural element. Though juveniles that had only lately emerged from the shell have been repeatedly met with, in no instance was I enabled to watch the female in the act of conveying her offspring from the nest to the water. On Fidra, a rocky island a few miles further up the Firth, I examined, in the summer of 1867, a nest containing four eggs that was placed on a shelf of the cliff at an elevation of forty or fifty feet above high-water mark, from which the young must have experienced some difficulty in making their escape. Within a distance of twenty yards a couple of nests had been constructed, under the shelter of a huge block of stone, in such close proximity that the down round their edges was intermixed.

On the Fern Islands, where I also inspected the breeding-haunts of these fine birds, they may be termed partially domesticated. The females appear to select the neighbourhood of the store-houses and other buildings as a protection from the attacks of the Lesser Black-backed Gulls; these robbers, whose quarters are near at hand, evince a decided partiality for the eggs of the Eider, and should an opportunity occur they

* These plants are probably *Beta maritima* and *Lavatera arborea* (sea tree-mallow). The fishermen from Canty Bay who act as guides usually inform inquisitive and credulous strangers that the latter flourishes only on the rock, bestowing upon it the name of Bass mallow.

wanting. The colours of the soft parts taken down on the spot may be thus described :—Beaks a dusky olive-green with a shade of yellow towards the point, nail lighter; legs and toes a dull greenish yellow, webs a dark grey; iris hazel.

The plumage of the adult males, I conclude from observations made on fresh-killed specimens, may be considered to be in the most perfect state about the end of January. So early as April the feathers show signs of wear, becoming by degrees more shabby till the general moult takes place towards the close of summer.

Large numbers of drakes frequented the Firth of Forth a few years back; in May 1864 and again in 1867 I repeatedly remarked flocks numbering from twenty to thirty, unaccompanied for the most part by females, making their way along the coast to the east of Dunbar. These birds usually flew in single file at the distance of about six feet apart, the long string of piebald fowl presenting a most singular and striking appearance. At this season, though their colours seem bright when viewed on wing, the plumage if closely examined will be found to have lost much of the gloss and beauty exhibited a few months earlier. During the latter end of May 1867, while staying at Canty Bay for the purpose of obtaining specimens in the adult summer plumage, I succeeded in procuring several drakes, and discovered that the whole of the primaries as well as the tail had faded to a rusty brown tint; the sickle-shaped secondaries were also much frayed at the points, in some instances the quills being perfectly bare. Several authors describe the inner elongated curved secondaries as white; this I have only found to be the case in males showing immature plumage; these feathers on adults in the most perfect state are pale buff with a sulphurous tinge.

The number of Eiders that resorted to this part of the coast twenty years back may be imagined by the following abridged remarks from my notes concerning the specimens described above.

Under the impression on the 25th of May (wind blowing strong from the east with a heavy sea rolling into the bay) that the flocks which had hitherto proved utterly unapproachable would be sheltering on the rocks and in the still water to the west of the island, I went afloat and proceeded up the Firth. In a small creek on the south side of Fidra a landing was effected at the second or third attempt; and by crawling over the slippery ledges of weed-grown rock I succeeded in reaching a point from which I was enabled to examine the western shore. At least thirty drakes, the majority in full plumage, were floating on the water, a few ducks and immature birds being also scattered here and there, while within ten yards of my place of concealment a dozen old males rested quietly in fancied security, with heads turned over on their backs. After watching the assemblage for some time, I secured sufficient specimens to meet all requirements in three shots, and further slaughter being unnecessary, Eiders were unmolested for the remainder of the day. The same birds in all probability were again encountered to the west of Ebris; another party, however, consisting of over twenty females and one male, rose on our approach, and after wheeling for a time over the breakers, settled among the rocks on the shore to the east of Gullane. Many other chances for procuring specimens were offered by birds feeding along the sands; and I have not the slightest hesitation in stating that, had there been any necessity, at least fifty Eiders might have been shot before we finally made the land. Unable to return by sea to Canty Bay, owing to the force of the wind and easterly swell, we ran ashore in Gullane Bay and, procuring a cart, brought our craft home by road. On all subsequent visits to this part of the coast I remarked that Eiders had greatly decreased in numbers.

On several occasions while shooting in the Channel about Rye Bay and off Pevensey Level during the winter months of 1858 and the four following years, flocks of large fowl, which I failed to identify, owing to their excessive wildness, were sighted a few miles off the land. Owing to the remarks in several works in my possession concerning the distribution of this species, I remained in ignorance that the Eider was a regular autumn and winter visitor to the English Channel till on January the 8th, 1881, a coast-gunner (who had seen the birds in my collection) reported having met with a drake in immature plumage off Shoreham on the previous day. The fact that one or two specimens (immature males) were procured along the shore near Rottingdean within a week of this date left little doubt as to the truth of the statement. It was not till the 4th of October, 1882, when a flock of eight immature birds busily diving over a stony bank a short distance off the beach near Lancing allowed the boat to run down before a fresh northerly breeze, that I obtained specimens off the Sussex coast. The two birds (both drakes) secured on this occasion exhibited a curious and mottled state of plumage: one of these juveniles being figured in Plate II. a description is rendered unnecessary; the colours of the soft parts, however, were as follows:—Iris dark hazel; bill at the point a pale greenish yellow, gradually changing towards the base into a blue-slate tint, a small oblong patch of yellow showing close in front of the nostril. The nail pale greenish yellow, the colouring being fainter than the bill. The lower mandible blue; legs and toes a dull olive-yellow; webs and nails a dark grey, almost black. All these colours commence to change within an hour or two of the death of the bird, and it is only by means of a plate life-size that any idea of the tints of the beak and legs can be given. The only food they contained consisted of small crabs, mussels, and shrimps, with a quantity of fine grit, diminutive beach-stones, and a little weed.

Since this date I have frequently recognized Eiders off the coast of Sussex; they were generally found diving for food over some bank where mussels, crabs, and other shell-fish abounded, or resting quietly out at sea a few miles off the land in company with Velvet and Common Scoters. The latest entry in my notes referring to this species is under date of December 29, 1883; the weather at the time being thick, without a breath of air, and the water as smooth as glass, it was impossible to sail down upon the flock. These Eiders were seven in number, two of which showed a small amount of white on the head and back, the remainder of the party appearing to be females. Owing to the presence of about one hundred and fifty Scoters, whose actions were very unsettled, constantly flying round, alighting and rising again repeatedly, there was no chance for a shot, and I failed to secure a specimen.

Along the coast of Norfolk and Suffolk small flocks were also occasionally observed in the winter of 1881 and the following year. I have here met with them off Hasboro' and Caister, also in Yarmouth and Corton Roads, and again further south off Pakefield and Covehithe. Though for the most part alone, they were occasionally in company, over the banks that formed their feeding-grounds, with Common and Velvet Scoters, as well as Long-tailed Ducks and Scaup; on rising on wing, however, they invariably separated from the rest of the fowl*.

According to the observations I have been enabled to make, the food of this species consists of various kinds of small shell-fish and marine animalcules, seaweed and grit as well as stones appearing to be swallowed to assist digestion. Specimens shot in the Firth of Forth in May 1867 contained large quantities of mussel-shells with a few small winkles, also sand and weed.

Having never kept Eiders in confinement, I can offer no opinion as to the age at which the males assume the full adult plumage, though doubtless it is not earlier than their fifth year.

The following description of the Eider, discovered in an old history of the Western Isles, may possibly prove amusing, if not instructive; to the best of my knowledge it has not been quoted by any ornithological writer:—

"In this island † there is a rare species of bird, unknown to other regions, which is called Colcha ‡, little inferior in size to a Goose, all covered with down, and when it hatches it casts its feathers,

* In December 1883 I received word that two or three Eiders in immature plumage were lately obtained on Breydon Water, near Yarmouth. In one instance a bird had been noticed swimming in the river among the shipping and barges off the wharfs; in all probability wounds or the continued buffetings of a heavy gale while off the coast would account for such unusual familiarity.

† Lewis.

‡ Evidently from the Gaelic "Colcach."

leaving the whole body naked, after which they betake themselves to the sea, and are never seen again till the next spring. What is also singular in them, their feathers have no quill; but a fine light down without any hard point, and soft as wool, covers the whole body. It has a tuft on its head resembling that of a Peacock, and a train larger than that of a house-cock. The hen has not such ornament and beauty."

A few words with reference to the several stages of this species figured in the Plates, together with the localities in which they were procured, may not be out of place.

Plate I. A female and brood (probably about three weeks old), obtained between the islands of Fidra and Ebris, in the Firth of Forth, during the second week in June 1867.

Plate II. Male and female in plumage of first autumn: the former shot in the Channel off Lancing, in Sussex, early in October 1882; the latter in Gullane Bay, in the Firth of Forth, on the 10th of September, 1874.

Plate III. Adult male, obtained near the island of Fidra in May 1867, and a male showing one of the intermediate stages of plumage, killed near the same spot in September 1874.

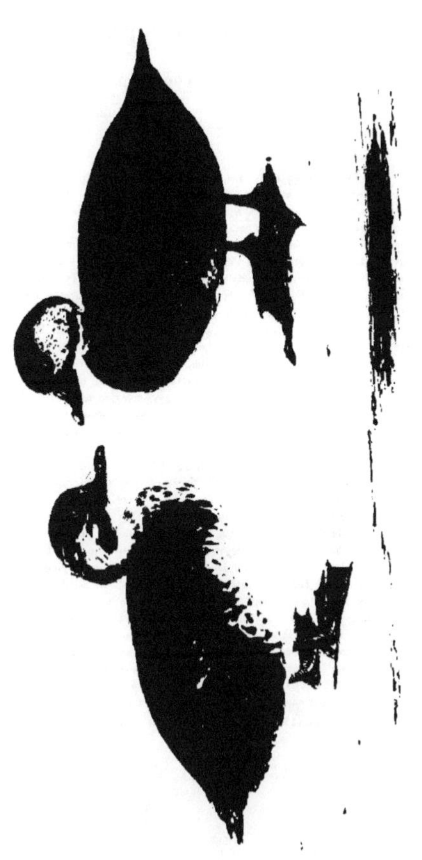

VELVET SCOTER.

ŒDEMIA FUSCA.

Large numbers of these fine birds are to be met with on the saltwater firths off the shores of Ross and Cromarty and the adjoining counties, as well as round other portions of the northern coast-line. On the Firth of Forth a few were now and then seen in Gullane Bay, and also off the Tyne sands near Dunbar, during the winters I passed in the locality. Off all suitable parts of the east coast these birds put in an appearance in larger or smaller numbers in autumn, remaining occasionally till the end of April or beginning of May. So far south as the Channel they commonly make their way by the first or second week in October, small parties regularly taking up their quarters by that date over the mussel-banks off Shoreham, Lancing, and Worthing *. These banks appear most attractive feeding-grounds for all the diving-ducks, immense flocks of Scoters of both species, together with a few Eider and Long-tail Ducks, gathering here in severe winters. In Rye Bay and off St. Leonards and Bexhill I also repeatedly fell in with Velvet Scoters when shooting in that part of the Channel. If hitherto unmolested, these birds are usually far more confiding than the Common Scoter, offering almost invariably an easy chance for securing any desirable specimens; a shot or two, however, puts them on their guard, when for the rest of the winter they will rise on the first signs of danger, spreading alarm to every flock of fowl in view. Though usually only consorting with diving-ducks, stress of weather will, at times, force this hardy species into less congenial society: on one occasion at least two hundred Velvet Scoters were seen in company with Mallard, Wigeon, Pintail, and various others of the web-footed tribe, a herd of Curlew, wading in the shoal water, being also intermixed with those nearest the shore. A few lines extracted and abridged from my notes for 1869 will give some idea of this immense gathering.

In the latter end of November a gale that swept over the north-east coast caused a heavy sea in the Dornoch Firth, and during the height of the storm thousands of fowl collected in the wash of the water, under the shelter of the sand-banks off the "Fendom." Having decided to make an attempt to reach them, we came to the conclusion that the single punt, owing to its lightness and the ease with which it could be dragged ashore, even if full of water, would be most suitable for the purpose. Owing to the swell, it was impossible to launch from the boat-house, so having procured the assistance of two or three extra hands, in addition to my regular crew, there was little difficulty in transporting the punt, the big gun, and other necessaries along the sands. Having made our way as near as possible to the fowl without giving any alarm, I succeeded, after shipping one or two seas while getting afloat, in working up within range, the rough weather having apparently rendered the whole assemblage heedless of danger. At last, with a couple of inches of water above the bottom boards, I paused within thirty or forty yards of the stragglers at the west end of the line to watch the actions of the fowl, till a glance to the north, where several breakers were following one another in rapid succession, warned me that if a shot was not instantly

* In 1880 an adult male was shot off Brighton on the 18th of August; this, however, was an unusually early arrival.

obtained all chances would be lost, as the punt must inevitably be swamped. So shouting to the best of my ability and flinging up an old sou'wester, I fired slightly above a dense mass of birds about fifty yards distant. That the attempt to rouse them had proved a failure was only too evident, not a fowl springing on wing till after the flash of the gun. The moment after pulling the trigger I seized a heavy shoulder-gun and left the punt to wash in towards the shore, making, as the mass of fowl broke up into endless flocks, a successful shot at eighty or a hundred Wigeon that drifted back nearly over my head. The depth of the water at this time of tide being scarcely two feet, there was not the slightest difficulty, unless an unusually heavy sea swept past, in wading towards the sands. The men who had followed my movements under cover of the banks now arrived on the scene, and proceeded at once to haul the water-logged craft ashore, and do their best to put gun and punt in working order. With the help of the retriever, who had joined me immediately after the shot, I secured fourteen Wigeon stopped by the shoulder-gun, as well as five more knocked down while flying round over the cripples. The pursuit of the wounded having led us some distance towards the east, I became aware that the charge from the big gun, though failing to take effect on the birds for which it was intended, had nevertheless reached those further off. A couple of disabled fowl (probably immature Long-tails) were diving and flopping with broken pinions out towards the firth, and "Nell," having made a vain but plucky attempt to reach them in the surf, fell in with a drake of the same species floating dead on the water, after turning to make her way back to the shore. While the men were completing their work I learned that a number of fowl some distance beyond those at which I fired had risen on wing, and were fluttering over the waves out towards the firth, the moment before the flash of the gun was seen. Two birds having the appearance of Geese were stated to have fallen to the shot, as well as one or two Curlew from a flock flying round at the time. On searching further east, an old Velvet Scoter was detected washing along the shore in a patch of weed, and "Nell" having been despatched to take a turn over the bent-grass on the sand-hills, brought back a screaming Curlew that must have run at least half a mile from where its wound had been received. During a break in the squalls another Scoter was observed paddling slowly out to sea over the swell about a couple of hundred yards out in the firth; these were doubtless the birds that one of the men had mistaken for Brent Geese, the drifting rain and scud having rendered it by no means easy to ascertain the identity of fowl at any distance. Though the punt-gun, a four-bore, and two barrels of a ten-double had been discharged, I only succeeded in bagging nineteen Wigeon, a Curlew, a Velvet Scoter, and a Long-tailed Drake, the two last being remarkably fine and well-marked birds; had the shot from the big gun not proved a failure, it is hard to say what number might not have been obtained. At the very lowest computation, I should think that four or five thousand fowl were gathered along the shore over a space of little more than a quarter of a mile. It was easy to identify Mallard, Wigeon, Pintail, Scaup, Velvet Scoters, and Long-tails; the Curlew, I was informed by the men, had been resting huddled up in the shoal water, till rising on wing in response to the call of a small party of their own species passing over. As no varieties had come under observation, and much time and labour would be needed to rectify the effects of the thorough drenching that all the gear had undergone, I started back for Tain, leaving the men to bring on the punt and heavy guns. On reaching home I learned that kindly neighbours had twice brought word of the sinking of my punt in the firth. To any one, however, who did not fear the effects of a saltwater bath there was not the slightest danger in the attempt to obtain a shot. Though the chances of success had been exceedingly small, it was a most interesting sight to watch the thousands of fowl quietly riding over the swell or spreading their wings for a moment to escape the effect of the broken water. When a lull occurred, I remarked that, with but few exceptions, the Wigeon and other fowl within sight turned their heads over on their backs, being doubtless weary and worn out by the long-continued buffetings of the storm.

The upper mandible of this handsome species exhibits a striking combination of tints. The colours on the soft parts of an adult male shot on the Dornoch Firth, off Golspie, on the 14th of April, 1869, may be described as follows:—The protuberance on the upper mandible is black, the colour stretching just below the nostrils, both in front and at the sides; the centre down to the nail is white; the sides (the colouring stretching almost to the base, which is black for a little over an eighth of an inch) are of a pale yellow, with a strong apricot tinge towards the point, edged on the lower portion with a narrow though clearly defined black line, the nail showing a dull yellow tint. Though the coloured plates in two or three of the most trustworthy ornithological works give a strongly-marked black line down each side of the white patch on the centre of the upper mandible, I have never observed this mark on fresh-killed specimens obtained either in spring, summer, or winter; possibly it does not make its appearance till shortly after death *. Iris silvery white. Legs and toes cherry-colour, inner side of legs orange, webs black; joints dusky; hind toe variegated cherry and black outside, inside an orange tinge. The colours of the soft parts of an immature male obtained off Shoreham in December 1879 were as follows:—Upper mandible a dark brown with lightish spots at sides, lower a dull pale brown. Iris dark hazel. Legs and toes a dull copper-colour, webs dark brown.

Though unable to state from personal observation that this species rears its young within the limits of the British Islands, I have little doubt that Velvet Scoters, in former years, regularly nested in two, if not three, of the northern counties of the Highlands; whether they still continue to do so, I can, however, offer no opinion. While collecting in the north in 1869, I learned from a forester, whose knowledge of the habits of the birds frequenting his native district was unquestionable, that the large Scoter † bred on the moors surrounding a number of small lochs in an open part of the country. My time being fully occupied in procuring specimens of several species whose whereabouts had been previously ascertained, I sent one of my men, who was also well acquainted with this Scoter, to visit the locality in company with the Highlander, in order to make sure no mistake had occurred. On their return I learned that a pair or two of birds had been observed, and a female identified on rising from her eggs. As arrangements had been made to leave on the following day for a drive of nearly one hundred miles, it was too late to put off our journey to visit the spot. Since that date I have not had another opportunity of exploring the district, though in a still more desolate and remote corner of the Highlands I watched a female flying over the moors, near a small swampy loch, one night in the summer of 1878; it is, however, quite probable that on some future occasion a downy brood of Velvet Scoters may be met with in one or other of the northern counties of the Highlands.

* A sketch I made of the head of the male described above (shot April 14th, 1869), taken immediately after the bird was lifted into the punt, corresponded precisely with that of the male figured in Mr. Gould's carefully executed plate of this species, with the exception of the absence of this black line, and the fact that the nail extended somewhat more prominently beyond the point of the beak. If I remember rightly (though no reference is made to the subject in my notes), this portion of the mandible contracted slightly in course of time.

† The male, as my informant described, much resembling a Blackcock when on wing.

SCOTER.

ŒDEMIA NIGRA.

At all seasons the Scoter is to be found on the seas surrounding the British Islands, most numerous in autumn, winter, and early spring, though a few stragglers remain in small parties all through the summer. The earlier writers on ornithology failed in most instances to note the breeding of this species in Great Britain; many, however, nest on the flat moors round the lochs in the more open portions of the country in Sutherland and Caithness. I have also received word from keepers that a few pairs resorted to some small lochs in the north-west of Ross-shire all through June and July; doubtless these were rearing their young in this district, though the fact escaped observation.

The Scoter, according to my own experience, is a late breeder; it is not till the first or, more often, the second week in May that the birds appear at their haunts on the inland lochs, though flying visits may have been paid to the locality earlier in the season. By the end of May I have observed as many as five or six pairs resorting almost constantly to the piece of water they had selected, having evidently taken up their summer-quarters. In early morning the drakes may be seen paying their respects to their mates, and in a sportive and most amusing manner flapping round and round the object of their admiration; after this performance is ended they wash, dash, and splash in the water, sending the spray flying in all directions. As soon as the females have commenced to sit closely, which usually happens about the second or third week in June, the males are less frequently seen making repeated flights to larger lochs or the open sea. The nests are placed in heather of moderate growth on the open moors, seldom within the distance of a quarter of a mile of the lochs; I met with a few, however, on the small peaty islets among the black pools on the flocs. Unless the old bird is driven off or watched to her quarters the exact position of the nest is by no means easily ascertained. The eggs are a dull dirty yellow, the lining of their cradle being composed of a plentiful supply of black down from the breast of the female; the young on leaving the shell are covered with a dusky down, the legs and beaks exhibiting a smoky tint.

During spring I have often watched immense flocks in the channel off Rye Bay and Fairlight going through much the same performances indulged in on the Highland lochs; the only difference was that the amusements seemed more general, numbers chasing one another while flapping over the surface, diving and ducking incessantly, the excitement being carried on with but slight intermission for an hour at a time, the surface of the water for a quarter of a mile or more appearing at times as if caught by a sudden squall, the black plumage of the birds as they sprang above the foam contrasting strongly with the spray dashed up, and forming a most singular sight. Though almost invariably watchful, Scoters now and then while so engaged permit a near approach; on one or two occasions I have worked the boat within fifteen or twenty yards of the nearest of these wideawake fowl. At dusk in spring there is also a general meeting in various parts of the Channel; at this hour they are generally far more noisy, their whistling cry being heard on all sides of a fine still evening as the darkness closes in. In such numbers do these birds collect in the vicinity of the mussel-banks about Goring and off other parts of the coast of west Sussex in winter and early spring, that I have not unfrequently heard the fishermen, while discoursing on the subject in answer to inquiries, remark how many acres of Coots * they had seen.

During January and February 1882 I passed several days in steaming up and down off the Norfolk and Suffolk coast from Hasborough to Southwold, in order to study the habits and actions of the various species of Ducks resorting to the salt water off these sandy shores, and having mounted a breech-loading punt-gun on the bows there was little difficulty in procuring any specimens required. Scoters were met with over the whole distance, straggling parties of from a score to three or four times that number being generally seen about Horsey, Winterton, Caister, and Yarmouth Roads, while off Corton, Pakefield, and Covehithe, and still further south, they were not unfrequently observed in flocks of from three to five hundred strong. As I was anxious to procure specimens showing the various stages of plumage, several shots were fired and some curiously marked young males obtained. When first approached the birds took little notice of the steam-boat; the same flock would, however, seldom permit a second chance for inspecting their ranks till several hours had passed. On one occasion (January 21st) we steamed slowly towards about three hundred busily feeding over a bank within a quarter of a mile of the sandy beach at Pakefield; when little over a hundred yards distant the whole body dived, and putting on full speed we reached the spot where they had disappeared before a single bird had risen to the surface. For at least fifty or sixty yards round the vessel the water seemed alive with birds as they popped up almost as instantaneously as they went under. A few again ducked down in their alarm, but the greater number sprung at once on wing, affording an excellent chance for securing with the shoulder-gun two or three exhibiting a curious mottled plumage. Had slaughter been desirable, and the steamboat eased or stopped about sixty or seventy yards sooner, a heavy shot might have been obtained; to kill more than were required, however, was useless, as this species, like the rest of the family, is utterly unfit for the table, its flesh being rank and unpalatable. On the 11th of February, while steaming towards the north, a shot was fired from the big gun about half a mile off Winterton; a fresh breeze was blowing from the north-east, and the crew of a yawl stationed at the look-out near the beach evidently mistook the sound for a signal from one of the floating light-ships. In less than a minute sixty or seventy men were busily engaged in running their craft down the sands, and a strong crew having been taken on board, they put off at once, evidently bound for the 'Cockle,' our attempts to induce them to turn back being utterly disregarded. On the 14th immense numbers of Scoters, together with flocks of Velvet Scoters and Long-tailed Ducks, as well as a few Scaup Ducks, were met with scattered over the roads between Yarmouth harbour and Corton Church; not needing specimens, ample opportunities were afforded for closely examining the various parties. The peculiar form of a short-tailed bird, however, attracted attention, and this oddity I was enabled to procure with the shoulder-gun, causing but little alarm to any of the surrounding fowl. On lifting the unknown to the deck in the deidle the species was readily ascertained, viz. a Common Scoter, in all probability suffering from wounds; the wings and the plumage on the head, neck, and portions of the back were much the same as those of immature males, though somewhat lighter, while the only covering on the rest of the body was a thick down mottled with grey and brown, the absence of the tail-feathers rendering its appearance especially strange.

There are usually every winter large flocks of Scoters harbouring at sea just off the boat-sheds in which my punts are housed on the shingle-banks near Shoreham; a few birds, usually old males, often separate from the main bodies, which generally keep about a mile at sea, and singly or in small parties of two or three fly in towards the shore and alighting outside the breakers dive busily for food. On the 10th of January, 1881, I happened to turn the glasses on to one that approached within the distance of eighty or one hundred yards, and remarked that, though apparently a fully adult male, the bird showed

* In several parts of Sussex the name of Coot is given by the seafaring population to the Common Scoter.

a conspicuous white patch on the throat and neck. After feeding some time, he paddled out to sea, and placing his head over on his back, drifted slowly off with the tide; a punt was then run out, and the water being as smooth as glass, there was little difficulty in sculling within range for the shoulder-gun. The bird on being picked up proved to be in particularly bright glossy plumage, and in addition to the white patch on the throat and neck, the colouring of the beak, on which the knob was especially well developed, differed considerably from the usual form. The orange mark on the upper mandible was continued down to the point including the nail, while the whole of the lower mandible was also a deep yellow. As this bird was perfectly crammed with small crabs, it is probably in quest of these dainties that the Scoters so frequently fish along shore. Some years back in a large flock off Bexhill I clearly identified an old male Scoter with a perfectly white breast and belly; this singular bird was seen repeatedly, but I failed to obtain a shot. A pair somewhat similar were observed the following year; but the distance at which they were examined was too great to allow a positive assertion as to the species.

In winter Scoters not unfrequently appear in small flocks on the Norfolk broads and meres: on one occasion I noticed nearly twenty flying round Hickling Broad in summer; these were doubtless non-breeding birds, resorting to the "Wold," that had taken an unusually extended inland flight. While gunning on several of the freshwater lochs in the North-eastern Highlands during the winters of 1868 and 1869 I did not observe this species, though Scaup and Tufted Ducks, with which they often consort, were exceedingly numerous.

A few remarks concerning the soft parts of the specimens procured while making observations on this species off the Norfolk and Suffolk coast in January and February 1882 may not be out of place. Two drakes in perfectly adult and most glossy plumage differed considerably in the form and colouring of the beak.

No. 1. An adult male with protuberance on upper mandible exceedingly prominent. The coloured line down the centre a very deep orange, almost vermilion, deep orange round nostrils, the lower part of the coloured patch being of a rich chrome-yellow. The remainder of the mandible with the protuberance a jet-black. Dark orange circle round eye, which in all specimens examined was a dark hazel. Legs, toes, and webs black.

No. 2. A male with the protuberance considerably smaller; the yellow line down the upper mandible exceedingly faint in parts, merely a small star showing above the knob, the colouring being also paler and the lower portion almost a lemon-yellow. The ring round the eye chrome-yellow. Iris, legs, and feet corresponding with No. 1.

No. 3. An adult female. Upper mandible a dull slate-black, straight, with no knob*, a flesh tint showing inside the nostril. Legs and toes a dusky olive-yellow, webs a dull brown and the joints clouded.

No. 4. An immature male, no protuberance on upper mandible. The coloured patch much the same in shape as on adult male, though the upper portion was somewhat broader. The colouring pale yellow, with an orange tinge round nostrils. No ring round eye. Legs and feet same as adult female No. 3. The plumage of this bird was a dull grey-brown, with a few black feathers showing here and there in patches.

No. 5. An immature male, with slight protuberance on upper mandible. The coloured patch was merely a pale yellow streak between the nostrils and an oblong patch of the same colour lower down, the latter mottled with a few small black spots. Inside of nostrils yellow. No circle round eye. Legs and feet same as adult female. A uniform dull dark brown, with few conspicuous markings, pervaded the whole of the plumage.

The immature males figured on the Plate are the two specimens (4 and 5) referred to above. As to whether they exhibit the plumage of the first or second winter, I should be sorry to venture an opinion: the age at which this species assumes the perfect adult dress seems also extremely doubtful.

* I remarked a slight protuberance on the upper mandibles of two females obtained in one of the northern counties of the Highlands during the breeding-season of 1869.

POCHARD.

FULIGULA FERINA.

THE range of the Pochard extends over the British Islands from north to south: in the east it is exceedingly common in all suitable localities; in the west, however, though the species is said to be by no means scarce, few have come under my observation.

The shallow lochs in the eastern portions of Sutherland and Ross-shire are visited in autumn by immense flocks of these divers, the majority usually taking their departure during March, the date varying somewhat according to the weather. At times, when in small parties, they may be seen in company with Scaup and Tufted Ducks, the larger bodies more frequently remaining unmixed with strangers. Loch Slyn, a mile or two inland from the Dornoch Firth, formerly possessed great attractions for these birds, the flocks, when disturbed, leaving the water and betaking themselves to the firth, returning usually in less than an hour to their favourite quarters. Though exceedingly unsuspicious on their first arrival, they speedily acquired the knowledge that the presence of a punt boded no good; and no sooner was the craft afloat, than spreading out into a long line they swam slowly towards the east, ultimately, if pressed, getting on wing and making across the sand-banks towards the firth.

On the east end of Loch Shin in Sutherland an excellent chance of a heavy shot at the most compact body of Pochard I ever had the luck to approach within easy range was accidentally lost at the moment when a mishap appeared impossible. Shortly before daybreak on the 9th of March, 1869, I was afloat waiting for the Dun-birds usually met with on this part of the water. Having at length, soon after sunrise, sighted the flock, numbering at least three hundred, about the centre of the narrow end of the loch, the punt was sculled cautiously towards them, the fowl at the same time opening out in line and paddling rapidly away towards the east end. A cart just then came in view along the road which borders the loch-side, and the whole body of fowl turned slowly round, evidently undecided what course to follow. At the very moment the birds had collected into a dense mass within sixty yards of the muzzle of the gun, and while merely waiting for the cart to pass out of the line of fire, the driver turned the horse's head to the water and pulled up in order to allow the animal to drink. The sudden stoppage caused the unsuspecting flock at once to face the punt and warned them of the impending danger. While getting on wing, and again after turning to make their way up the loch, they offered most tempting chances; but the proximity of the hotel and neighbouring houses rendered it utterly impossible to fire with safety at any elevation above the water. For the remainder of the day the birds proved restless; and moving the punt further north on the 10th, it was not till the 23rd of the month that I was again on the water. The greater part of the Pochard and the whole of the Tufted Ducks had now taken their departure, and little sport was met with during the day. As the night was fine and still, with a good moon, I started soon after 11 P.M., having previously despatched a man or two to drive some small pieces of water among the hills to which the birds frequently resorted at feeding-time. While waiting quietly under the shelter of the bank on the south shore of the loch, the bows of the boat just touching the stones, a white hare came slowly down from the moor to the water-side. After a few minutes she approached the punt and smelling it carefully, hopped right on to the deck. Here she appeared perfectly at home, turning her head from side to side and regarding us with the greatest unconcern. Having indulged at length in a sniff at the muzzle of the big gun, the scent of the powder evidently proved distasteful, and a rapid retreat was made to the bank. Still puzzled by the aspect of the punt and its crew, she refused to quit the spot; and when at length we moved further west, she accompanied us on the loch-side, stopping from time to time to gaze intently from the bank while the craft continued in view.

Pochards, on their first arrival on our northern shores in autumn, are by no means shy, but are gradually rendered excessively wary by the constant persecution to which they are exposed in almost every county while on the journey southward; till in the end no little skill is needed to work a punt within gun-shot by day. The large flocks may, however, be approached with some chance of success just before daybreak. After feeding during the night they seem disinclined to get on wing with their accustomed promptness, drawing at first together if carefully manœuvred; occasionally, however, they spread out in line and swim rapidly away, when, aided by their largely-webbed and powerful feet, pursuit is utterly hopeless. I had made a heavy shot at these birds some winters back in the east of Norfolk, and the following morning was again on the look-out for the flock. Though prevented by a thick fog from finding them for a considerable time, I at last caught sight through the haze of apparently one hundred at least swimming close together at a distance of about sixty yards, and had brought the gun to bear—my finger was on the lanyard—when suddenly a head surmounted by a sou'wester was raised in the centre of the object, and the flock of Pochard resolved itself into a man in a punt, lying broadside within ten or fifteen yards of the muzzle of my gun. The imperfect light, coupled with the dense easterly haze, imparted to the long low craft the exact outline presented by a string of these fowl when sunk deep in the water to avoid danger.

There are no cripples so difficult to gather as the diving ducks, and the Pochard is certainly one of the toughest to disable outright. After the big gun has been successfully fired, and a goodly number of dead and wounded sighted on the water, the work is far from over: spreading out in every direction the broken-winged birds rapidly separate, and with the least breeze on the water, it is highly improbable that even a quarter will be recovered. If carefully watched during a dead calm, it will be seen that on coming to the surface for air in many instances a portion only of the beak is exposed, the bird again instantly submerging itself with scarcely a ripple to betray its position. At times, with head and neck extended level with the water and flapping vigorously the uninjured joints of the wings, they will rush for some twenty or thirty yards over the surface, then disappearing with a sudden spring, vanish entirely from view. On the lochs of the north and the broads of the eastern counties where the water is preserved there is, however, little difficulty in eventually securing the greater number of the wounded, as on searching round the reed-beds or among the water-plants in still weather they may generally be detected. A frost sufficiently hard to "lay the whole of the bush" * at once drives them to the open water, when, deprived of all cover, they fall easy victims.

In the eastern counties Pochards exhibit a decided partiality to the society of Coots; it is seldom the two species are seen on the same broad unless in company. Some twelve or fifteen years ago, I often watched immense bodies feeding in company, two or three acres of water having been at times almost black with birds; their numbers, however, unless alarmed and put in motion, could scarcely be estimated, as many were constantly disappearing below the surface.

Previous to leaving the Norfolk waters on the approach of spring, I have repeatedly remarked small

* In the east of Norfolk the reed-beds are invariably termed "the bush." When a piece of water is frozen over, it is said by the natives to be "laid."

parties composed entirely of drakes in the brightest stages of plumage. For a week or ten days after the 19th of March, 1873 (the weather at the time being cold with strong north-easterly breezes), I noticed that the birds, though previously wild, were now exceedingly confiding, allowing the punt to approach within the distance of fifty or sixty yards without exhibiting the slightest signs of alarm. In order to ascertain whether they had acquired the rank flavour usually looked for in wildfowl when the shooting-season is over, I took advantage of an unsuspecting half-dozen floating quietly with their beaks buried in the plumage of the back, and securing the whole, discovered that for the table they had by no means deteriorated, though Mallard and Wigeon were at this date decidedly unpalatable. The Pochard, according to my opinion, is the only diving duck that can possibly lay claim to be considered a delicacy; and even at its best, after feeding regularly on the inland broads or lakes, this species is not to be compared with any of the small-footed wildfowl, such as Wigeon, Pintail, Shoveller, or Teal.

While afloat on the saltwater firths of the north, or the muddy harbours and estuaries of the southern and eastern coasts, Pochards are forced to subsist on minute shell-fish and insects of various kinds, as well as a certain amount of marine vegetation. Their favourite diet, however, consists of portions of freshwater weeds or plants that grow so profusely in the shallow lakes and pools to which this species would doubtless resort more frequently if unmolested. Formerly when decoys were more common in the eastern counties, and even since the pipes have ceased working, immense flocks of Pochards rested during the day on the protected water, making their way as soon as darkness set in for the large broads, in order to make a meal off their favourite weed. The old gunners in several districts speak with unfeigned regret of the immense flocks that came "roaring" over certain spots in the marshes regularly at dusk, affording in squally weather, when the line of flight was low, excellent chances for picking up a few couple of birds. Of late years, since most of the decoys have been done away with, the numbers of Pochards have greatly decreased in the district.

Though I failed to detect the nest, and never met with a chance of examining the young in down, it is well known that Pochards breed in more than one locality in Great Britain. A few stragglers continue in the Highlands during April, but after that date the species is seldom observed. On two or three occasions, however, between the 11th and 20th of June, 1869, when visiting Loch Slyn, I noticed a couple of drakes on the water: these birds appeared unusually regardless of danger; and anxious to ascertain whether they remained of their own free will, or their presence at this season was the result of wounds, I pressed them closely in the punt, when rising at once on wing they made the circuit of the loch, and after a short flight returned to the water. Judging by their actions, I am of opinion that the females were sitting on some of the marshy spots near at hand.

My notes for the summer of 1873 contain a reference to a flock observed in Norfolk in July; a short extract may not be out of place, as the state of the weather is given, as well as a list of other strangers putting in an appearance at the same time:—

"July 26. Not a breath of air below, though the clouds were moving from the south-west. Soon after daybreak there was every appearance of a tempest, the sky to the north and west being as black as night. I had often previously remarked during stormy weather at this season that flocks of waders, as well as sea-birds and fowl, were noticed in the district. There were this morning on the various hills round the broad numerous small parties of Curlew, Whimbrel, Greenshank, Golden Plover, Dunlin, and Ring-Dotterel. An immature Green Sandpiper was shot, and also a couple of Black Terns, the latter exhibiting the change into their winter dress, a state of plumage in which this species is seldom met with. A bunch of about a dozen Pochards continued flying over the water for some time; they seemed inclined to settle, but were at length disturbed by a shot. Before mid-day the weather cleared off, and the whole of the strangers had taken their departure, the hills being deserted save by the resident parties of Redlegs, Peewits, and Snipes.

From the view obtained through the glasses at the Pochards, it appeared that none showed the perfect adult plumage."

Though the autumn flocks seldom arrive before October, an old male or two may occasionally be recognized about the Norfolk broads all through summer. On the 1st of September, 1871, while stationed in a punt on the upper part of Breydon, a drake flew at a considerable height direct over my head, and on being knocked down by a shot from a shoulder-gun, proved to be in very richly coloured, though worn and faded, plumage *.

Though the larger bodies pay little or no attention beyond wheeling round once or twice, small parties of Pochard almost invariably alight close to the wooden decoy-ducks I make use of in winter. Ten or a dozen carefully constructed dummies moored by a ring in the breast to a small anchor so closely resemble the wild birds as they ride head to wind, rising and falling with the swell, that it is almost impossible to scare away a fowl or two of this species that have been attracted and settled in their company. On the 20th of January, 1872, by way of ascertaining how long they would remain, I repeatedly drove up with the punt a couple of fine drakes. The poor birds continued swimming round their inanimate companions, turning and regarding the group intently when the boat approached within the distance of ten or a dozen yards. Though occasionally rising on wing, they resolutely refused to quit the spot, returning after a short and hurried flight, and in the end I left them floating contentedly in the midst of the fascinating dummies.

In the eastern counties these birds are usually known to the natives as Pokers or Sandy-head Pokers †, Dun-birds or Redheads being the titles most commonly assigned to them by the fowlers along the south coast.

Paget's Pochard.—Scientific naturalists are, I believe, of opinion that this form is merely a cross between the Common Pochard and the Ferruginous or White-eyed Duck. After comparing the specimen from which the figure in the Plate is taken with the two species, it was obvious that there could be little, if any, doubt on the subject. Unfortunately no opportunities fell in my way for acquiring any information concerning the habits of this fowl, and I am only enabled to give extracts from my notes for 1871 concerning its capture, coupled with the state of the weather at the time of the occurrence.

November 13. Hard frost, light north-easterly breeze. On reaching the water-side a couple of hours before daybreak, the punts were discovered frozen firmly in, a thick coating of ice having formed in the dykes. While the men were clearing the punts, I made my way across the hill towards the broad, the moon shining brightly and lighting up an exceedingly wintry scene. The whole of the south side of the broad appeared laid with thin ice, with the exception of a small piece of open water near the point of the hill, in which a number of Mallard were feeding and slushing about, evidently enjoying themselves to the fullest extent. Well aware that it was utterly impossible, owing to the ice, to approach the party in the punt, I endeavoured to stalk within range from the bank before the noise of the boats crashing up the dyke should disturb them. All went well until within about sixty yards, when a crust of ice over which I was crawling gave way and instantly put up the fowl. On the arrival of the punt, we broke straight through to the open water, and then edged round the ice to the western side of the broad, in order to take advantage of the first signs of daylight. Shortly after reaching the outskirts of the large reed-bed, a small bunch of fowl (evidently Pochards)

* A somewhat singular mishap occurred at the shot. Owing to the gun (a heavy muzzle-loading 10-bore, carrying in the two barrels a quarter of a pound of shot) having been accidentally forgotten while exposed to a passing shower, the first barrel hung fire, and on the second trigger being pulled, both charges exploded at the same moment. Though no recoil was noticed (merely a a slight numbness being experienced in the fingers) the gun flew over my shoulder, and was discovered with the barrels firmly fixed in the mud between three and four paces from the boat.

† Col. Hawker states (eleventh edition, page 203) that "the *dun-birds* are called *red-heads* on the south and west coasts, and *Parkers* or *half-birds* in the fens." Parkers, I conclude, must be a misprint for Pokers, the well-known east-country name, never having heard the former title made use of in either the fen or broad districts. Goldeneyes and Tufted Ducks are occasionally termed half-fowl by the east-coast gunners; I cannot, however, recall to mind having heard this name applied to Pochards.

came in view, paddling slowly in line. There were not above half a dozen in all, and for a time a more favourable chance was awaited. As the light increased, it became evident that with the exception of a single bird, I was unable to identify, feeding round a patch of floating weed at the distance of about fifty yards to the right, no other fowl could be reached without breaking through a quantity of drift-ice. At last the Pochards came slowly together, and dropping the barrel of the big gun, the lanyard was pulled; a snap, however, was the sole response. Though not the slightest attention was paid to the crack of the cap by the party at which the gun had been aimed, the single fowl rose instantly on wing, and flew straight to cross the muzzle of the gun. A second later the charge exploded, the shot being fatal to the half-dozen Pochard as well as the unknown, which fell stone-dead only fifty yards ahead of the punt. The first rays of the sun were now breaking through, and as the bird was lifted on board, the singular bloom of claret tint on the head and neck was distinctly visible and at once attracted attention, leading eventually to the discovery of the white eye, dark back, and light markings on the wings.

An elaborate description of the specimen is not required, as the Plate will supply all necessary information concerning the plumage.

SCAUP.

FULIGULA MARILA.

Flocks of these hardy birds are to be found off almost every portion of our coast-line: though by no means so abundant of late years, I used in former days to meet with immense numbers, during severe winters, diving for food over the sandy flats off the shores of Kent and Sussex. In Rye Bay, where this species is known to the local gunners as the "Frosty-back Wigeon," the whole of the fowl were much cut up by the hard weather in December 1859, and from a fishing-smack furnished with a long gun secured by a breeching in the bows, I obtained several shots at even the largest flocks of Scaup and Wigeon. These birds seldom make their way inland in the south of England; few, if any, were ever noticed on the lakes or rivers at a distance from the coast, and the visitors to the harbours or brackish pools inside the shingle-banks were almost invariably cripples that had escaped the shore-shooters. The Norfolk broads, so attractive to Pochards, appear to offer but few inducements to this species to prolong their visits when driven in by stress of weather or other causes; a party numbering from ten or a dozen up to a score may now and then be seen on the water, but a move to other quarters is usually made before many hours have elapsed. While collecting specimens in the east of Ross-shire and Sutherland in 1868 and the following year, I often met with these birds on the inland lochs within a few miles of the coast; they were also repeatedly observed on the saltwater firths, flocks numbering from one to two hundred birds at times affording good sport while punt-gunning.

In the Dornoch Firth numbers of these birds used to resort at high water to the banks between Tain and Morangie, where food might be obtained, occasionally making their way as far up as Edderton Bay. Unless in want of specimens or to procure some fowl for any of the country people who had made application, I seldom molested them, as Scaups, in my opinion, are far from palatable when prepared for the table. The largest flocks proved exceedingly wild and much averse to allow a near approach when an attempt was made to scull within range for the big gun; under sail, however, with a fresh breeze, it was easy to run down within forty or fifty yards before a bird would rise on wing. On springing from the water the line of fowl, which in most instances had previously spread out, would at once draw closer, and an excellent chance was not unfrequently offered. After the manner of all the large-footed diving Ducks, the cripples are difficult to recover when a successful shot has been obtained. Spreading out in all directions and moving rapidly under water, they are speedily out of sight; with the slightest ripple on the surface pursuit is almost useless.

On the 19th of April, 1877, while driving from Inverness towards the south, along the shores of the firth, I noticed a score at least of Scaups in company with Mallard, Wigeon, Goldeneyes, and Mergansers, all in magnificent plumage and exceedingly tame, swimming in detached parties, and feeding here and there among the weed-grown blocks of stone within a short distance of the road. The tide was high at the time, and the birds were plainly visible without the aid of glasses: as they spread out while searching for food, it was evident that the greater number, if not the whole, were paired, the ducks and drakes in every instance keeping company, and one or the other with croaking notes occasionally resenting a near approach from another couple.

After many attempts to ascertain the truth, I am unable to state with certainty that Scaups rear their young in this country; though failing to detect their nests after most careful watching, I remarked that a few pairs occasionally remained all through the summer in the Highlands. I often noticed two old drake Scaups, late in the spring of 1869, on Loch Slyn, in the east of Ross-shire; they also resorted to a small muddy rush-grown piece of water, with which the loch is connected by a narrow stream of about a mile in length, just navigable for a single punt. The swamps and stunted plantations, rank with moist undergrowth, around this paradise for birds were frequented by many kinds of fowl; and I have little doubt that the nests of more than one species, supposed to be only winter migrants to our shores, might in those days have been detected, had sufficient time been devoted to the search. During the course of a thorough exploration of this locality in the first week in June, I distinctly saw a female Scaup crossing the marshes on wing, though the drakes, as far as I was able to ascertain, were absent; on the following day this bird, or another of the same species, was seen on the water in the early morning and again towards evening.

While travelling by rail to Fearn, on a visit to the haunts of the Scaup about Loch Slyn, I was much amused at the manner in which one of the fishwives was observed to provide for the safety of her railway ticket while on her rounds. The train had just pulled up at a small station, when a bright-looking lassie, in the familiar garb of the fishing-population, alighted, and lowering her well-laden creel to the platform, proceeded to divide her return ticket into the two halves. She next lifted one foot on to the top of the basket, and raised her dark blue skirt; then turning down her stocking, she carefully enveloped one half of the ticket in two or three folds taken round the top. Having completed this little arrangement to her satisfaction, she swung the dripping basket over her back, and turning towards the outlet-door, delivered up to the collector the requisite portion of the ticket, and started off to dispose of her load of haddies.

TUFTED DUCK.

FULIGULA CRISTATA.

This hardy diving-duck is to be found in every part of the British Islands I have visited, from north to south, usually resorting to fresh or brackish water when in quest of food, the lochs of the Highlands, the lakes and reservoirs of the midland as well as the broads of the eastern counties forming its favourite resorts; to the pools in the shingle-banks and the estuaries intersecting the mudflats in Kent and Sussex it is also occasionally driven in severe weather during the winter gales. The Tufted Duck has little beyond the handsome plumage of the drake when in full winter dress to recommend it, as for the table this species, whether old or young, is rank and unpalatable.

While staying for the punt-gunning at Tain, on the shores of the Dornoch Firth, in the winters of 1868 and 1869, I remarked that the flocks of Tufted Ducks took their departure about the end of March from Lochs Shin and Slyn as well as the other large pieces of fresh water on which they were usually to be found after the beginning of November.

Early in November 1871 these birds, for the most part in immature plumage, were exceedingly numerous on Hickling Broad and Heigham Sounds in the east of Norfolk. Not needing any, unless fine old drakes as specimens, I had refrained from molesting them, till at length about half a dozen that happened to be intermixed with Pochards and Coots were stopped by a shot with the punt-gun on the Sounds. While collecting the cripples a mishap occurred which, though merely causing slight damage to the punt, might have been attended by serious consequences. Having fired the big gun, I sculled up to collect the slain, and on arriving within range of several lively cripples, a 12-bore breech-loader was made use of and speedily put a stop to all attempts to escape. The whole, with the exception of one or two that dived off towards the reed-beds, having been accounted for, the shoulder-gun was recharged and laid down so as to be within reach of the right hand while working the craft. My attention was then turned to collect those floating around; while picking up three or four drifted together under the bows of the punt, the retriever (until now curled up under the stern deck) drew up, and by some means, probably through her stepping on the triggers, one barrel of the 12-bore was discharged, the shot passing under my arm as I knelt on the bottom boards, and blowing a round hole in the side of the punt. As I had left my station at the butt of the gun and crawled up to the beam of the fore deck in order to reach the birds, there was little blame to be attached to the poor old creature*, who doubtless imagined she was merely doing her duty in following my movements. Since this occurrence I have, however, never taken a dog afloat in a gunning-punt when working up to fowl, and fully intend to follow the same rule for the future. While on Hickling Broad on the 7th of November, 1871, I sighted a drake in handsome plumage in company with half a dozen immature birds. Being anxious to secure the adult male, the punt-gun was fired, and the whole party turned over, apparently dead, to the shot. On nearing the spot,

* The retriever referred to was "Nell," frequently mentioned in the pages of 'Rough Notes.'

however, the drake showed signs of life, and after a vain attempt to rise disappeared below the surface. As I saw he was hard hit and the water was without a ripple and as clear as glass, escape was evidently impossible, so taking a 12-bore I stood up and awaited his appearance in order to administer the quietus. After watching for a minute or two some bubbles were noticed coming to the surface, and on shoving up to the spot, I could plainly see him at the depth of about three or four feet struggling violently with his head held down as if in a noose. In a few moments he was quiet, hanging like a criminal, only in the inverse manner, caught by the neck. While skimming along close to the weeds at the bottom his head must have become entangled with some of the long twining strands, and unable to free himself he had naturally succumbed to drowning.

The old drakes of this species do not, as a rule, arrive in any numbers on the Norfolk waters till severe weather has set in; on the 4th of February, 1873, several were observed during the day on the open parts of Hickling Broad and also on the Sounds. While making our way slowly homewards in the punts through the broken ice, as darkness was setting in, a large flock of black and white Pokers * were heard passing overhead. Judging their whereabouts by the sound of their noisily beating pinions, I fired both barrels of a heavy 10-bore and a couple were heard to crash down on to the ice. One bird had fallen within a few yards of the boats, and being dimly visible on the rough snow-crusted ice, was easily reached and proved to be a remarkably perfect Tufted drake. The second fowl appearing to have dropped into some thick cover near a small island, was left for the night, and on reaching the spot shortly after daybreak on the following morning a Goldeneye drake in full plumage was discovered lying on the ice among the reeds.

During the intense cold in January 1881, several flocks of Tufted Ducks came under my observation, passing along the Sussex coast, about half or a quarter of a mile at sea off Shoreham and Lancing, making their way towards the west. Small parties also and single birds frequented the harbour and the lower pools in the river Adur, alighting when a chance was offered, though but little rest could be obtained, owing to the number of gunners lining the banks adjoining the towing-path. The food to be obtained in the large pools of brackish water inside the shingle-banks near Lancing were also a great attraction; though repeatedly fired at and driven out to sea, they were sure to return and settle down again after having swept round the spot to ascertain whether any danger was to be apprehended. On the 15th and again on the 17th, the day preceding the terrible gale and snow-storm that caused such destruction to bird-life, I was afloat in the gunning-punt, and remarked that although some hundreds of this species must have passed, only a single male in full plumage came in view, flying straight out to sea from the harbour, apparently disgusted with the inhospitable reception met with inland. So severe was the frost on the days we were out in the Channel, that the Rangoon oil used to clean the action of the big gun froze in cakes on the metal, and the pin fitting into the socket of the sliding-knee had to be kept in working-order by a constant application of paraffine. A few showing the various curious stages of immature plumage were shot about the pools and backwaters in the salt-marshes during the first few days after the storm; these, however, merely indicated the usual changes gone through before arriving at maturity and need no description in these pages.

While shooting along the coast in the west of Sussex in the winter of 1874, I explored the estuaries and creeks about Bosham and Emsworth, and also the channels leading to Chichester harbour, and remarked that the Tufted Duck was spoken of by the puntmen and shore-gunners of the district as the Covey Don.

* Goldeneyes and Tufted Ducks often join in company when on the Norfolk Broads, and are generally considered by the local gunners and marshmen to belong to one and the same species, the birds composing the various flocks being each and all known among the shooting community of the district as black and white Pokers.

GOLDENEYE.

CLANGULA GLAUCION.

Though much time was spent in the attempt, I never succeeded in verifying the fact of this species breeding in Great Britain. The female I repeatedly observed during summer on remote Highland lochs, and on more than one occasion a bird was detected flying from old and weather-beaten pine-woods, where doubtless her nest was concealed; sweeping rapidly beneath the branches in the shade thrown by the dense timber, it was by no means easy to keep the small grey-tinted fowl in view. For several days subsequent to the 11th of June, 1869, I observed a male, at times in company with a female, disporting himself on the waters at the east end of Loch Slyn in Ross-shire. While watching the pair the female on one occasion disappeared without attracting attention; shortly after, however, she came in sight, skimming from the plantation of Scotch firs standing to the east of the loch, and rejoined the drake, when both birds, evidently disturbed by the approach of the punt, rose on wing and left the water. The male evinced a decided aversion to permit of a close inspection; judging, however, from the nearest view I was enabled to obtain, he appeared to be in full adult plumage. This is the only instance where the mature drake came under my notice later than the first week in April.

This species is common during autumn, winter, and spring all round the Scotch coasts, as well as on the inland lochs. The numbers decrease somewhat towards the southern counties of England, though I have occasionally known several small parties and single birds to remain for some weeks about the muddy harbours and estuaries of the Sussex and Hampshire coasts. I find by my notes for 1868 that Goldeneyes had arrived and taken up their quarters along the east coasts of Ross-shire by the 22nd of October, a few stragglers having been noticed a week or ten days earlier. There is very little difference in the date at which these birds reach the Norfolk waters; and though seldom seen so early off the Sussex coast, I remarked a pair in immature plumage resorting almost constantly to the pool in Marazion Marsh, near Penzance, from the 23rd to the 29th of October, 1880.

Compared with the number of immature birds and females, the full-plumaged drakes are exceedingly scarce in all parts of the country; it is seldom they are observed earlier than the middle or latter end of November. But few of these attractive birds are driven to our shores till severe weather accompanied by protracted frosts has set in; occasionally in open winters I have failed to notice a single specimen in perfect adult dress, though numbers in the various intermediate stages might be met with.

The immature drakes frequently remain late on our coasts; on April 1st, 1869, while gunning on the Dornoch Firth, I fell in with a small party of half a dozen Goldeneyes near Morangie. After firing the big gun, I discovered that the whole were males, the dead and wounded consisting of one adult and five immature, the latter all in much the same stages of plumage. On the 28th of April, 1873, a pair (an immature male and female) were shot on Hickling Broad; these birds rose from the bank of one of the hills on which they were resting when alarmed by the punt. On no other occasion have I fallen in with this species on land unless disabled by wounds. Another pair, also male and female, in immature plumage, were seen for three or four days after the 8th of May, 1883, on Hickling Broad; the former exhibited the white patch on the brown head most conspicuously.

With the exception of the handsome old males which are invariably sought after, punt-shooters have little or no inducement to molest this species, its market value being exceedingly low. For the table it is utterly useless, the nature of the food consumed (minute shell-fish of various kinds, as well as the innumerable insects that inhabit both fresh, brackish, and salt water) rendering the flesh exceedingly strong and fishy *.

In order to procure specimens in various stages of plumage, I frequently worked up to and examined the parties of Goldeneyes met with while gunning on the northern and eastern coasts. Though the adult males will be found the most wary of the web-footed tribe, the younger birds usually fall easy victims, for the most part evincing little distrust till repeatedly alarmed. When wounded few fowl are able to make more rapid progress through the water; aided by their large and powerful feet and small though sturdy pinions, they prove exceedingly hard to recover. Towards the end of December 1881 an immature male I was anxious to secure was lost on Hickling Broad, after having been disabled by a shot, and on the following morning a careful search was instituted round the edge of the hills. At last the bird was detected squatted on the side of a marsh-wall, my attention being first attracted by the brightness of the pale yellow eye in the midst of a tuft of dead thistles.

To attempt to give all the local names by which the Goldeneye is known to the gunners on various parts of the coast is utterly impossible. In many districts the females and young of several of the diving fowl are supposed to belong to one species, and necessarily much confusion exists. In the east of Norfolk I have usually heard the immature birds termed black and white Pokers, while the adult males are honoured with the title of "Old Hardweathers." Goldeneyes when on wing may be heard for some considerable distance, the sharp beat of their strong flight-feathers producing a sound that has led to the names of Whistler and Rattlewings being bestowed on the species.

While shooting in the east of Ross-shire it was a rare occurrence to pass through the upper waters of the Dornoch Firth without finding a fine old drake or two about some of the larger pools between Bonar Bridge and Inveran. Having procured a few as specimens, I noticed it was seldom that their favourite quarters were long deserted, the spots vacated by those killed being speedily filled by fresh comers. During the early part of the winter of 1868 I had made repeated attempts to obtain a shot at an adult pair that frequented the Kyle below Invershin; the vigilance of the drake, however, enabled them to evade all dangers for a time. Early on the morning of the 9th of December, while making our way down towards the firth, a thick haze was encountered, drifting slowly in from the east. Suddenly through a break in the mist a couple of fowl were detected diving busily close to the south shore in the large pool above Bonar Bridge; favoured by the shelter of a point of land extending some distance into the water, the punt was worked fairly within range of the big gun before an alarm was raised. On rising simultaneously to the surface after a lengthened dive the birds sprung at once on wing, and crossing each other in the act of turning, a successful shot was obtained, the pair falling dead within a yard or two of the bank. For a few moments the smoke mingling with the fog hung heavily in wreaths in front of the gun, and it was not till nearing the fowl that a sound resembling a low groan caught my ear, and at once drew attention to the shore, where a horse, apparently dying, was discerned through the haze. On landing, the true state of affairs was soon ascertained; the animal, doubtless swept away and drowned in the upper part of the glen by a recent flood, had eventually stranded at the entrance to the pool; by the side of the carcass, and now in his last gasp, lay an unfortunate collie, that had evidently been tearing at the flesh when struck down by the shot. At the first glance in the imperfect light the

* The stomach of a female Goldeneye shot 11th of January, 1883, contained at least a handful of small black bugs of about half an inch in length, also a quantity of vegetable remains, as well as gravel and sand. The insects were probably captured in brackish water.

struggles of the dog had given rise to the impression that the horse had fallen at the discharge of the gun. The fowl, on examination, proved to be the adult pair so frequently observed resorting to the pool; a week later I remarked that this favourite station was again tenanted by birds in the same stage of plumage.

An adult male in perfect plumage is figured on the Plate; the specimen from which the sketch is taken was shot on Heigham Sounds, in the east of Norfolk, in March 1873. An immature male, killed on Hickling Broad in April 1873, is also shown; this bird exhibits a state of plumage that is probably assumed after the first moult and carried for one or two years. I have had no opportunities of ascertaining from personal observation the age at which the male puts on the full adult dress. The white patch on the brown head of the immature male appears to have escaped the notice of most writers. MacGillivray states that "young males are distinguishable by their greater size and darker tints from juvenile females." While progressing towards maturity he says that they have "white feathers intermixed with black before the eyes:" this remark probably applies to the white patch in its earliest stage, but is scarcely descriptive of the appearance of a young male at the close of the first winter. Dresser alludes to a young male as resembling the female, but with rather more white on the wings. It is strange that this distinction should not have been observed by such an experienced sportsman as Charles St. John, who remarks, when referring to this species, "both female and young male are without the white spot under the eye."

SMEW.

MERGUS ALBELLUS.

The Smew appears to be met with more frequently on the southern and eastern coasts of England than on any other portion of the British Islands; during the seasons I passed shooting and collecting on the firths of the Northern Highlands and about the sandy bays of the Forth in East Lothian, not a single specimen was met with.

In the winters of 1859 and the three or four following years several, in various stages of plumage, were obtained by the local fowlers about the open-water dykes and pools in the marshes round Rye and Winchelsea, as well as on the river and harbour near the former town. An old gunner who had followed his calling with the help of an antiquated and rickety musket, apparently utterly unserviceable, but making, according to its owner's report, most marvellous shots, brought for my inspection an adult male in the finest state of plumage; this handsome fowl he persisted in calling a "Greenland ice-bird." I did not then succeed in securing a single specimen, either adult or immature; a duck and drake knocked down with the same barrel one wintry evening in December 1861, while flying out of the Nook, were carried out of the harbour on the ebb-tide, the rapidly increasing darkness precluding all chances of falling in with them if followed in a boat. A few days later, on a cold frosty morning shortly after daybreak, I was watching another fine old drake flying down from the upper part of the river in a line for the harbour, and remarked that the course he was holding must bring him within range, when a couple of shots fired by a gunner concealed in an old and disabled craft lying on the mudflats, though apparently with little effect, caused the bird to swerve to the west and lower his line of flight. As he gradually dropped I took out the glasses and kept him in view, till, while crossing the line of railway towards the harbour mouth, he came in contact with a notice-board, and after fluttering a short distance, rose a few feet in the air and instantly fell. On making my way to the spot, I mounted the post of a marsh-gate to obtain a view, and at once detected the old drake with his wings spread out in a small drain, his conspicuous plumage at once attracting attention. Having a good idea as to where the gunner who fired the shot might be found, I picked up the bird and, after returning to the harbour mouth, embarked in a small flat-bottomed boat in which I had been accustomed to explore the shallow creeks of the Nook, and took the flood-tide up the river shortly after midday. After leaving the water it was a long tramp across the marshes; but taking advantage of one or two inviting slades, a few couple of Snipe were bagged, and the shots eventually attracted the man of whom I was in search. Much surprised at the sight of his Smew, which he fully believed had carried off the charge, his thanks were profuse; and in return I enjoyed many a good day's sport at the long-bills when in the district, some of the slades and creeks in this locality being about the most attractive feeding-grounds and harbour for Snipe that have come under my observation.

During the severe weather succeeding the terrible gale of the 17th of January, 1881, I noticed several Smews in immature plumage in the Channel off Shoreham, Lancing, and Worthing; on the 22nd and 28th they were especially abundant, and a few were shot by the shore-gunners about the creeks and in the harbours. On several occasions I attempted to scull up within range; but the birds were invariably on the alert, and to approach within the distance of one hundred yards was utterly impossible; running down before the wind under sail was also tried with a like result. One adult male only, so far as I was able to ascertain, put in an appearance in the district; in company with a female, he was discovered feeding in the basin at Southwick by an inexperienced shooter, who succeeded in stalking within a fair distance, and, considering the bird with a red head a more desirable specimen, slaughtered the female with a steady pot-shot and allowed the drake to escape without making an attempt to use his second barrel. The weather during the latter end of 1881 was unusually mild, westerly and southerly winds prevailing, and but few of the diving-fowl were seen, though I kept a constant look-out on one or two of the larger broads in the east of Norfolk. Dense sea-fogs with scarcely a breath of air set in early in 1882, and for weeks the screeching and bellowing of the fog-horns and whistles on the light-ships off the coast were a constant infliction; such a continuation of open weather necessarily accounted for the small number of foreign fowl on our waters. On Hickling Broad I noticed but one immature Smew and an adult male Goosander; the former was shot, while the latter, after resorting to the open portion of the water for a few days, took his departure unmolested, evidently possessing a strong objection to permit a near approach.

The young male depicted in the Plate was obtained on Hickling Broad on the 24th of January, 1882; when first seen, his small size at once attracting attention, the bird was flying in company with five immature Goldeneyes. After wheeling round several times the party settled on the open water and commenced plunging below the surface; the presence, however, of the other watchful Divers rendered it impossible to obtain a shot at the tiny stranger: after being twice alarmed they rose high in the air and left the water, heading straight for Horsey Mere. A couple of hours later a small white-breasted fowl was observed diving close under the dark shadow thrown by the muddy bank of one of the hills, and an inspection through the glasses instantly revealed the fact that our friend of the morning had returned alone, and was busily engaged in searching for food. Not a breath of air was stirring, and the water being as smooth as glass, little time was lost in sculling towards the spot; having arrived within the distance of fifty yards without causing the slightest alarm, we paused to watch the manner in which he pursued his prey. Several times this active little Diver returned to the surface, having evidently met with no success, as after looking wistfully round he instantly plunged again. At last with an unusual flutter, causing a perceptible ripple on the water, the hungry bird dashed up to within thirty yards of the punt, making frantic but apparently vain attempts to swallow a fish protruding at least a couple of inches from his bill. With distended throat and widely opened mandibles, he swam round and round in circles, stretching forward his neck, and repeatedly dipping his bill below the surface for the distance of two or three yards; from time to time he lifted his head in the air with a resolute shake. After these antics had been continued for some five or six minutes, the bird seemed to have satisfactorily disposed of his troublesome capture, and rising half out of the water commenced flapping his wings in the most vigorous manner; this was a chance not to be lost, and a charge from the punt-gun at about sixty yards laid this diminutive wanderer from the north dead as a stone on the water. On examination the stomach was found to contain one fresh roach of such dimensions that, when the small gullet of the bird was considered, it appeared a mystery how the little glutton succeeded in getting it down. There were also the bones of another fish of the same species, several minute shells and stones, and some fibrous grassy roots; the latter were probably portions of the weed torn up from the bottom when the shells were swallowed.

The colours of the soft parts were as follows:—Iris very dark hazel; mandibles, upper and lower, slate, the line of the teeth a lighter hue, nail also lighter. Inside of mouth and tongue slate, changing into a flesh-tint towards the point of the beak. Legs and toes slate, webs darker, with a narrow line of a lighter tint each side of the toes.

RED-BREASTED MERGANSER.

MERGUS SERRATOR.

The Red-breasted Merganser is by far the most numerous of the family of sharp-billed Diving-Ducks that frequent the British Islands. This species may be met with in small flocks, during most winters, along the sea-coast of the southern and eastern counties. I have not recognized it in the north of England on more than one occasion, though in the south of Scotland it appears less scarce. In the more northern parts of the Highlands this handsome bird is a resident, breeding in several counties and on the adjacent islands.

Between fifteen and twenty years ago small parties were occasionally to be observed about the channels in the muddy harbours of Chichester, Emsworth, and Bosham; and in December 1859 I killed a remarkably fine male at Pagham Harbour *, near Bognor. The rosy hue on the breast of this specimen was deeper in tint than in any bird I have subsequently obtained or observed. Numbers showing the plumage of the female or the various immature stages may also be seen annually in flocks of from half a dozen to fifteen or twenty off the coast between Littlehampton and Shoreham. These birds for the most part resort to the open sea, though when undisturbed they not unfrequently make their way to the small pools of brackish water inside the sea-beach, and occasionally proceed some distance up the rivers. This species is not so plentiful towards the east of Sussex; but a few are to be met with along the coast off Pevensey Marsh and also in Rye Bay. Some years back, before the water was drawn off from the pools and channels about the harbour and marshes round Rye, these spots were frequently visited. The constant persecution, however, that they met with from the numerous gunners eventually drove them from this district, before the reclaiming of the land rendered their old haunts unsuitable. On the last occasion (January 1866) that I passed a week for fowling in the bay and marshes, I did not observe more than one or two single birds, and these stragglers kept entirely to the salt water.

Along the Norfolk and Suffolk coast these birds are usually to be seen during autumn, winter, and early spring. The salt-water harbours about Lowestoft and Yarmouth provide a supply of suitable food, but the professional gunners are far too watchful to allow them many chances of settling. Off this coast I have never observed the Merganser so plentiful as in the English Channel. In severe weather a few may be met with on any open pieces of water on the rivers and broads of East Norfolk. According to my own experience in this locality, the Merganser is not so plentiful as the Goosander. Old males in perfect plumage, as is the case in all parts of the south, are exceedingly scarce as well as wary. During hard frosts in the winters of 1860 and 1861, I met with this species on several occasions in open parts of the Cam, near Cambridge, and in the neighbourhood of Waterbeach and Wicken fens, also more than once on the backwater near Granchester. These birds did not frequent the marshy pools on the poor lands or fens, resorting only to the rivers or lodes †.

Along the coast of East Lothian flocks of Mergansers appear early in the winter; the estuary of the Tyne, near Dunbar, is a favourite spot. At times they make their way to the pools on the links; but in such

* This harbour, a first-rate resort for fowl in days gone by, has now been drained for some years.

† Local name for a broad water-dyke, capable of navigation by flat-bottomed boats.

cases the absence of fish is speedily detected, and a short visit the result. On two or three occasions I observed a male and female in company remaining in this locality as late as May; though frequently watching their movements when feeding in the shallow water round the rocks, and also at the mouth of the small river, I could detect no signs that they were nesting in the district.

On almost every part of the coast-line round the Northern Highlands the Merganser is a constant resident. The breeding-haunts are usually not at any great distance from salt water. Along the western shores the numberless rocky islets and the wild and rugged country, intersected with hundreds of small lochs in the vicinity of the coast, afford suitable nesting-quarters. On the east the birds not unfrequently penetrate to more remote inland lochs.

About Gairloch, on the west of Ross-shire, Mergansers are exceedingly plentiful, and being seldom interfered with, become as tame as farmyard Ducks. I have repeatedly watched two or three females fishing in a small burn running down within ten yards of the front door of the inn*. The party swam in one by one from the bay with the flood, and, joining in company about the top of high water, made their way beneath the bridge, where they eagerly searched the small pools for fish. These birds, as far as I could judge, were sitting on the islands in the bay, and regularly left their nests at this time of tide to seek for food along the shore. The males were seldom seen near at hand, though one would occasionally join the flock if they got on wing after being disturbed. As the females swam back towards the islands where their nests were concealed, a drake would occasionally accompany his mate a short distance, but invariably got on wing before the foot of the rocks was reached. I have more than once had the glasses on the spot to watch the bird make her way to the nest after having observed her leave the water; but although well aware of the exact position of the eggs, she invariably managed to regain her quarters unobserved. In some instances I detected a regular track beneath the heather, along which she was enabled to creep without attracting attention. At this season (the latter end of May) the males had lost the fine plumage they exhibited in the winter and early spring; the green of the head was now clouded with brown, the broad white ring round the neck had disappeared, and the red or tawny feathers on the breast had assumed a more dusky hue.

On Loch Shin and other large lakes in the east of Sutherland I often remarked a somewhat singular habit of this species. Two, three, or at times even four broods join in company immediately after leaving the nest. Occasionally all the females may be observed with this school of youngsters, but, as a rule, only one is seen †. Having noticed these birds repeatedly making their way back on wing from the direction of the coast, it is probable they betake themselves to the salt-water firths or the open sea. During the height of summer I never observed a Merganser showing any thing approaching the male plumage. I conclude they either entirely assume the dress of the female, or retire for a time from their usual haunts.

The nest is usually carefully concealed beneath an overhanging slab of rock, or in some slight cavity in a heathery bank; at times, on the shores of inland lochs, I have seen them well hidden among thick bushes and shrubs, or under the stem of some rugged and weather-beaten tree. The eggs, most frequently from eight to ten in number, are warmly covered with the grey down from the breast of the female. The outer portion of the nest is composed of dried strands of coarse grass, with occasionally a few small heather-stalks.

The nature of their food, which consists of any small fish they can capture, sand-eels, shrimps, and the various insects of repulsive aspect that inhabit both fresh and salt water, necessarily imparts a strong and disagreeable flavour to the meat; and I can safely assert that the Red-breasted Merganser is utterly useless for the table. On different parts of the coast I have heard these Divers spoken of by the local gunners as Sawbills or Spear-drakes. When unmolested, there are few birds so confiding; though if once their

* The erection of the new hotel has doubtless altered all this.

† This curious custom often gives rise to the idea that the family of the Merganser is more numerous than it is in reality.

suspicions are aroused by a shot or two, I am unacquainted with any of our wildfowl so utterly unapproachable on open water.

Plate I. shows the female with a newly-hatched brood shortly after leaving the nest. I came unexpectedly on a family-party sunning themselves on the rough and stony bank of a loch in the east of Sutherland, and noticed the whole group collected together in the manner depicted by the artist. This occurred on June 29, 1868. A week or ten days later I again found other broods in the same stage.

In Plate II. a male shot in Gairloch on May 20, 1868, is figured. In many instances I remarked that the males have commenced assuming this dingy-coloured plumage as early as the middle of March. On April 13, 1869, I met with from one hundred and fifty to a couple of hundred of this species swimming in small parties on the still water in the upper part of the Little Ferry, a muddy salt-water loch near Golspie*. Never having been disturbed, the birds were by no means wild, and the punt was sculled slowly through the numerous parties while I carefully noted every stage of plumage. The two brightest specimens that came under observation were obtained, and I discovered they closely resembled the bird depicted in the Plate. This large gathering may be accounted for by the birds having collected along the coast previous to dispersing to their various breeding-quarters.

The full plumage of the adult male is given in Plate III. The specimen from which the sketch is taken was shot in January 1869 on the Dornoch Firth. This bird had attracted my attention by his conspicuous colouring for several days. The small flock with which he kept company usually resorted to the waters of the firth between Edderton Bay and Tain Sands. Several attempts had been made one frosty morning during the flood-tide (the navigation of the firth being at the time somewhat difficult, owing to drift-ice) to obtain a shot at the party. The presence of the wary old drake had, however, in every instance rendered our efforts to get within range an utter failure. Shortly before the turn of the tide, as a number of immense blocks of ice were banked together in a most threatening manner a short distance up the river, it was considered safest to draw in towards the landing-stage at the Meikle Ferry, and knock off work for a time till the full strength of the ebb was spent. In those days my boats were housed in a shed built up against the inn; and while the men were dragging one of our craft under cover, I remained on the small quay, watching the progress of a squall gathering towards the north-east, and keeping a look-out for any passing fowl. Just as the flakes of snow were commencing to fall, I happened to catch sight of half a dozen Mergansers settling on a large slab of ice about a couple of hundred yards to the west of the point. A momentary glance was sufficient to ascertain the fact that our old friend was among the number; so jumping into the double punt, which was still moored alongside the quay, I shoved off for one more trial. Unfortunately there was no time to summon assistance, as the birds were now rapidly approaching, and I was consequently forced to start short-handed. Luckily a current carried the ice on which the birds were resting over towards the south shore; and after clearing the cover from the lock of the gun, I had merely to make half a dozen strokes with the sculling-oar, and then crawl forward and draw the trigger, as the barrel was pointing fairly on the unsuspecting party. The old male at once turned over, and after giving two or three flaps with his wings, which carried him within a yard of the edge of the ice, dropped perfectly dead. It was but the work of a minute to run up alongside and, grappling the slab with the ice-hook, to seize my prize. I had no chance to collect another couple or two that were lying a few feet further on the ice; so after putting a stop to the struggles of one unfortunate cripple with the charge from a shoulder-gun, I turned my attention to the ice, which was commencing to move towards the sea with greater rapidity. It was perfectly obvious when I shoved off that there could be no chance of regaining the ferry-point, and my object was to reach the shore on the south side of the firth at the earliest opportunity. A dense squall of snow was now passing over, shutting out the view on all sides beyond a distance of fifty yards, and, owing to the darkness, rendering especially formidable the aspect of the ponderous

* This piece of water is marked Loch Fleet in the maps, but in the district I never heard this name applied to it.

slabs of ice pressing one on another as they heaved up and down in the tide, silently for the most part, though an occasional ominous craunch gave evidence of their weight and the force with which they were carried. The shouts of the men who had been attracted by the shot were plainly audible on the shore; it was, however, utterly impossible for them to render the slightest assistance. For a few minutes longer I was forced to drop gently with the tide, in order to avoid contact with some unwieldly blocks of ice and frozen snow which were rolling round in the current. I had learned by previous experience that extreme caution is necessary under such circumstances. The recollection of a most exciting half-hour, when carried away by drift-ice while making a vain attempt to recover a winged Swan on the ebb-tide in Rye harbour, had by no means faded from my memory. At last I noticed two large pieces, caught by a back eddy, turn over and take the ground, arresting the crowding blocks for a moment. This was a chance not to be lost; and a few strokes of the sculling-oar enabled me to part company with the drifting masses and gain fleet water. The banks were soon reached; and what might have been a most unpleasant adventure luckily avoided *. A couple of shots brought the men to the spot; and after an hour's hard work, hauling, dragging, and lifting over obstacles, the punt was safe back at the ferry-point.

The figure in the background is taken from an immature specimen obtained near Shoreham the last week in December 1882. I find in my notes the following remarks concerning this bird:—"Legs and toes bright orange; webs dusky; nails dark horn-colour; upper mandible dusky horn-tint; nail black; yellow line along lower part of upper mandible; saw-marks or fringe red flesh-tint; lower mandible cerise; lower nail pale yellow; iris olive-yellow."

* I remember hearing an account, though the particulars have now slipped my memory, concerning an accident that many years ago befell the ferry-boat near the same spot. This somewhat unmanageable craft happened to get caught by drift-ice, and was carried down the firth towards the bar. In this case, I believe, the boat eventually came back on the flood after a long and perilous voyage.

GOOSANDER.

MERGUS MERGANSER.

In several of the wilder districts of the Highlands the Goosander passes the summer and rears its young, remaining, in many instances, throughout the year. The numbers that visit the rivers and estuaries of the English coast-line during winter are exceedingly uncertain, varying with the seasons—a continuation of frost not unfrequently causing a general movement towards the south, while in open weather scarcely a bird is to be seen between the Humber and the Channel.

The Goosander has been recorded by more than one writer as breeding in the Outer Hebrides; I failed, however, to identify the species during my visits to the outer islands, and the nature of the country appears to differ considerably from the localities to which the bird, for the most part, resorts in summer.

Throughout many of the wild rocky glens in the more remote districts of the Central Highlands, where old and rotten timber, consisting of pine and birch, still exists upon the hill-sides, the adult males may be seen in company with small parties of females and immature birds during winter. When viewed in these deep dark gorges, the water, inky black from the shadows thrown by the overhanging pines, rippling silently round the stones on which he rests, there is no more attractive bird than the male Goosander, his salmon-coloured breast, green head, and bright red feet being shown off to the fullest advantage by the gloomy character of the surroundings. As late as the end of April I have repeatedly observed males and females, adults and immature, still in flocks; there is, however, but little difficulty at this season in distinguishing those that are paired, as the drake is unremitting in his attentions and seldom strays to any distance from his mate. Early in May I remarked that the flocks, in most instances, broke up, many of the birds appearing to take their departure from the district, pairs only for a time being seen in company. By the end of the month the males were less frequently noticed, and shortly after that date disappeared entirely, the females being seen invariably alone with the broods after leading their downy offspring from the hill-sides down to the rivers or lochs. At this season the males doubtless undergo their annual change of plumage; having failed to detect them about their inland breeding-quarters or on any of the adjoining lochs, I am of opinion that they make their way to the saltwater firths, where singly or in small parties they pass the remainder of the summer. Early in July 1878 I noticed three birds of this species on the waters near the head of the Cromarty Firth; though making no attempt to rise on wing, they succeeded in diving out of range. The view obtained was too distant to allow of accurately discerning their state of plumage; the general tint of the feathers, however, appeared similar to that of the females, though one or two exhibited several dark patches on the rufous colouring of the head.

An adult female with her half-fledged brood resting quietly in the bright sunshine on the unruffled surface of one of the larger lochs presents a sight that would doubtless prove puzzling to one unacquainted with the habits of these singular birds. The female, ever on the alert for the first signs of danger, floats motionless with her head drawn back and beak resting on the feathers of the breast, the youngsters by whom she is surrounded appearing to vary in colour from a creamy salmon to a dull slate. One moment half or three fourths of the brood show up the former conspicuous tint, while shortly after a transformation takes place and the colours are reversed. A glance through a strong binocular at once solves the mystery, and reveals the half-fledged juveniles spreading themselves out to enjoy the warmth of the sun. From time to time a portion of the brood turn over on their backs, remaining often in this position for several seconds; the next minute a bird or two may be seen, each with one foot flapping in the air and paddling slowly round with the other; while engaged in these antics the bright colours of the underparts are clearly exposed to view.

Such an appearance would be presented by a brood from five to seven weeks old; at this age the head, neck, and back still retain a quantity of the brown-tinted down, while the breast and belly are thickly feathered, the plumage being of much the same creamy salmon hue as that of the adult male. I have repeatedly watched this performance in still sunny weather, and remarked that till the young arrive at the age of a month or five weeks the female seldom leads her brood onto the deep waters of either loch or river, keeping them in the shallows, where there is less danger of their falling victims to the pike. On some of the larger sheets of water I noticed broods of young numbering only four or five, the remainder of the family having in all probability been snapped up by these freshwater sharks *. On one occasion (July 13, 1878), shortly before midnight, I surprised a brood resting on a spit of sand on the shores of a large loch; picking our way quietly homewards by the water-side, no notice was taken of our approach till within a couple of paces, when with a tremendous splashing the whole party fluttered and paddled out from the shore.

The young birds figured in the Plate were hatched in one of the large pine-forests that stretch for miles along the mountain-sides in the Northern Highlands. After bringing her brood down to the river-side, the female kept them for a week or more about some small pools that afforded a shelter from the strength of the current; as they increased in size, she worked steadily down the course of the river, being usually seen about the shallows, where flat or sloping ledges of rock and stone enabled the young birds to leave the water without difficulty. Being anxious to secure specimens previous to the feathers showing on the head and back, the brood had been examined on several occasions, and at last, when about seven weeks old, I came to the conclusion that they were in the stage required. On searching for the family, consisting of the female and ten young, they were discovered resting on the stones in a wide and shallow reach of the river, where there was not the slightest chance of approaching within range without attracting attention; the birds were also much scattered, rendering it doubtful if more than one or two could have been obtained at a shot. From having frequently watched their actions, I was well aware that though, of course, incapable of flight, they could both swim and dive at a pace that rendered pursuit through the steep and rocky glen entirely out of question. As the party remained for several hours without changing their position, beyond rising at times and stretching their necks and wing-joints, I decided on attempting a drive. Having first taken up a position so as to command a narrow gorge where the river swept down in a succession of falls between high slabs of rock, a keeper was despatched to show himself below the brood, and, if possible, induce them to move up-stream. I had previously remarked that when bent on making their way in one particular direction, no amount of driving that was possible in such situations would force them against their will; consequently it was satisfactory to see the brood, one by one, drop off the ledges on which they were resting, and with heads up-stream, paddle slowly in front of the dog which the keeper had sent out into the river. On nearing the rapids, I noticed the birds were swimming with only their heads showing above the water, which in this part of the river dashed through the narrow channel between the rocks with tremendous force. A low whistle was frequently uttered; but whether this was the note of the old bird or the cry of the young, it was impossible to ascertain *. By diving and swimming they succeeded in working their way against the current at a most surprising pace; so rapid were their movements in ducking below the surface, that though well within range and keeping in a compact body, I was unable to catch a glimpse of more than two or three of their number at the same moment. Under such circumstances the result of the shot proved far more satisfactory than was anticipated, five young birds and the old female being stopped by the two barrels. The survivors at once turned, and fluttering and diving, as well as aided by the current, passed rapidly down the river, in spite of all efforts of the keeper and dog to turn them †. Some time was spent in securing the dead and wounded; and when at last we followed the course taken by the remainder of the brood, our search, which was continued for a couple of miles, proved a failure, with the exception that another disabled bird was discovered by the dog.

The plumage of the female exhibited little difference to that of specimens obtained in winter; the crest on the head was, however, scarcely so long and thick. The young birds showed a thin covering of feathers on the crown and fore part of the head, the back of the head and neck being still covered with long reddish down. The throat was a mixture of white down and fine pin-feathers; back long brown down. Feathers of a slate-grey had expanded on the wing-coverts. The white of the bar on the wings had made its appearance, though but very slight signs of the pinion-feathers could be detected. The feathers of the tail had sprouted to a considerable length, the breast and belly also being thickly feathered and of almost the same rich salmon tint as in the adult male. The upper mandible brown along top ridge, the lower portion, including the saw, being of a flesh tint, the lower mandible of a deep red flesh. Iris dusky yellow, with darker outside circle. Legs a dusky brown tint, darker at the knee-joint; toes pale orange; webs dusky brown.

This brood, which appeared to be the only one in the immediate neighbourhood, had been watched for about seven weeks, and during that time had moved down the course of the river for nearly ten miles. These birds seemed, from some unknown cause, exceedingly scarce this season, in former years three or four broods having been usually observed on the same stretch of water.

Thoughout the districts in which I met with Goosanders during the breeding-season, the females appeared in some instances to resort to situations for nesting-purposes at a considerable elevation on the hills. A cavity in a large and partially decayed birch was pointed out by a keeper as the spot from which some eggs (previously seen in his possession) had been taken. The old and weather-beaten stump was on the outskirts of a thicket of birch, fir, and alder stretching from a swamp up a steep brae, and within a mile of a loch on which I have repeatedly watched two or three broods. The tree was carefully examined, and I noticed that down from the breast of the bird was still clinging to the rotten wood; the general appearance also of the rubbish in the hollow left little doubt as to the truth of the statement. On more than one occasion I have been informed by keepers and gillies well acquainted with this species that they had met with broods on the bare and open moors following the course of some of the larger burns. Whether these had been hatched among the rocks and stones in the rugged gullies near at hand, or still higher on the mountain-side where dense patches of fir clothed the slopes in the more sheltered corries, it was impossible to form an opinion.

Goosanders are blessed with a strong healthy appetite, their visits at times proving exceedingly distasteful to the custodians of lakes and rivers. When wounded or alarmed, I have occasionally remarked that an immense quantity of fish was thrown up. After a shot with a punt-gun, some winters back, on Heigham Sounds in the east of Norfolk, at a number of these birds sitting with other fowl at the edge of a wake on the ice, scores of small rudd and roach were discovered lying on the surface where the flock had been resting. On the upper waters of the Lyon, in Perthshire, while concealed among the alders on the bank of the river, I watched, at the

* I never heard any sound uttered by the female and brood unless disturbed and driven, or when moving off of their own accord. The note is a low plaintive whistle, not unlike the cry of some young Hawks.

† Though these young birds had not a pinion-feather on their wing-joints, they appeared on rising after a dive to spring upwards and flap along the surface for at least a yard or two, striking up the water at the same time with their feet.

distance of only a few yards, eight or ten immature birds diving for food in
stones. At last the party appeared satisfied and paddled slowly to some ledg
intention of landing, when, offering a good chance, five were stopped with th
trout, all perfectly fresh, that were shaken from their throats would have mc
sized fish-creel.

When unmolested this species is by no means shy; in many of the Hi
resting on the stones by the river-side, within a short distance of the road, pay
traffic. The most utter disregard of danger, however, that I ever witnessed w
male and female, that frequented the shores of the Beauly Firth between Cl
in 1878. On several occasions, at high water, they fed among the weed-gro
heads when approached within fifty or sixty yards.

During the winters I passed on the Norfolk broads, but few adult males
last specimen seen in these parts being an exceedingly brightly coloured bi
Hickling for several days from the 30th of January, 1882. It is seldom th
locality unless frost has set in; on this occasion, however, the weather was op
having prevailed for over a week, the wind varying from south to east. The yo
are far from scarce during most winters, being decidedly more abundant on th
Mergansers. In this part of the country, and indeed all over England, these l
almost unapproachable by the constant persecution to which they are exposed.
Pagham and more lately at Rye and Shoreham, I met with this species; they
the Sussex harbours than on the east coast.

The males of this species do not assume the fully adult dress their first wi
I obtained, and carefully examined, exhibited but little difference in colouring
shot in the west of Perthshire in December 1866, may be regarded as
shown during the first or second winter; this specimen is figured on Plate I
the first winter; there is, however, a possibility that it may be a year older. A
pronounce the bird a female, though after a close inspection it will be seen tha
slightly darker, the crest is wanting, and the small white patch below the
smaller. This rufous tint on the head and neck also terminates with an edgi
the throat, and is wanting in the female. The salmon tint on the breast a
as well as more extended, less of the clouded grey markings showing round th
along the flanks. I can offer no opinion as to the age at which the drakes o
mature plumage; unless the birds are reared in confinement, and every chang
no chance of ascertaining the various stages through which they pass.

The capture of the female Goosander in the manner described in the
improbable; I, however, merely give the statements concerning the occurrenc
leaving my readers to form their own opinion.

While driving through a wild and desolate Highland glen, in the summer
to explore the country and make a few casts for the small red trout swarming
rocky burn that dashed alongside the rough hill-track, when heavy clouds gath
of an approaching storm. The mountain-tops were already lost in the drivin
commenced to fall, when at a turn in the road a bothy, erected for the acc
foresters, came in view. Regretting that at this season the shanty would,
I, however, pushed on, and was agreeably surprised to find on the spot a keeper
other business, had looked in to ascertain that all was secure. The man pr

having placed the pony and trap in the best shelter available, we made our way indoors to await the abatement of a succession of blinding squalls of sleet and rain now sweeping up the glen. Uncertainty always existing as to where a halt is likely to be made when travelling through unfrequented districts, my conveyance, as usual, contained an ample supply of creature comforts, and the delay caused by the continuation of the storm was but little heeded. Having thoroughly discussed the prospects of the coming season, I happened to inquire if Goosanders were ever seen in this part of the glen; while passing along the burnside, a short distance below the bothy, my attention had been attracted by feathers, evidently from the wing of this species, partially trodden into the soil, and this had led to the question. On several occasions during the many years passed by my informant in this district, I learned that he had observed the female with her brood working down the burn towards the large lochs in the vicinity of the river. After describing an amusing scene that occurred some years back, when the gillies in attendance on a fishing-party, at the request of the sportsmen, had made vain and ludicrous attempts to secure one or two specimens from a brood surprised in a rocky portion of the burn, he concluded by stating that the previous spring an old female had been captured in the adjoining room to that in which we sat. Having made further inquiries, and carefully examined the premises, I ascertained that the facts of the case were as follows:—The bothy consisted of two rooms, the larger of which was used as the kitchen, and contained among other articles a large wooden meal-chest. At the end of the season previous to the last, when the building was locked up for the winter, this chest still held a quantity of meal, and on entering the kitchen and unbolting the shutters the following year it was discovered that a rat had gnawed a hole through the wood (round as if cut with an auger) in order to reach the meal. Again, as usual at the conclusion of the shooting-season, the bothy was shut up for the winter, and a second time a quantity of meal remained in the chest. In order to put a stop to the depredations of rats, an ordinary spring trap was placed near the old hole (now stopped by a large bung of cork), and in this a female Goosander was lying dead when the room was opened in the spring. In no other manner could the bird have effected an entrance unless by means of the chimney; the upper part, originally constructed of rough-hewn timber, now charred and weather-beaten, somewhat resembled the cavity in an old and rotten stump. It can only be supposed that the female, while searching for nesting-quarters, by some strange mishap made her way into the chimney, and having fluttered downwards to the kitchen, had flapped round the room in her fruitless efforts to escape, till taken in the trap.

A few words with reference to the Plates may prove of service:—

Plate I. Female with brood about seven weeks old. The specimens from which the figures are taken were obtained in the Highlands in July 1878.

Plate II. Adult male, shot in the north of Scotland in March 1878; also a male in the plumage of the first or second winter, obtained in the west of Perthshire in December 1866.

GREAT CRESTED GREBE.

PODICEPS CRISTATUS.

The breeding-haunts of the Great Crested Grebe appear to be now restricted to a few of the central, southern, and eastern counties of England: considering the persecution that this curious bird has undergone from one cause or another, it is a wonder that survivors are still to be found. Though its quarters are rapidly becoming contracted by drainage and other innovations, the broads in the eastern counties with their extensive reed-beds are likely to afford a safe asylum for many years to come; but the demand for plumes in former days, and latterly the rage for egg-collecting, coupled with the scanty protection afforded by proprietors in many districts, render it, unlikely that the species will increase. During winter, even should the weather prove mild, the majority of these birds take their departure from their summer-haunts and resort, for the most part singly, to the tidal rivers or the open sea. On the lochs and rivers of the Highlands I did not recognize a single specimen; though during a residence of two years in East Lothian a few pairs and solitary birds were not unfrequently met with off Gullane, and also in Aberlady Bay in the Firth of Forth.

The old birds doubtless take their young into the water almost immediately after hatching; on one occasion only have I met with a juvenile on the nest, the fact being referred to in the following extracts from my notes:—"June 26, 1871. Started early for the broad, to examine a nest discovered a few days before, containing four eggs, apparently on the point of hatching. On reaching the spot, it was ascertained that only two were left, the empty shells plainly indicating that a couple of the juveniles had already taken their departure. Though doubtless near at hand, it was useless to search for the youngsters, owing to a strong breeze that ruffled the water and kept the reeds in constant motion.

"June 27. Down at the Loons'* nest at daybreak. The two eggs were still unhatched, though one helpless mite was calling distinctly inside the partially broken shell. As an old bird, accompanied by the juveniles that had already taken their departure, could be heard close by, the patch of reeds in which they had taken refuge was surrounded by three punts; and the party having been driven backwards and forwards, the young were at length secured in a landing-net, the old birds following them to within a couple of yards of the boat's side. As soon as the capture was effected we retired from the neighbourhood of the nest, and took up our quarters in an adjoining reed-bed, so that the bird might return to hatch the remaining eggs. Some three or four hours later, having discussed our breakfast and indulged in a rest, the spot was again carefully approached: so noiselessly did the punt glide through the roadway cut through the reeds, that the female did not dive from her nest till the craft was within half a dozen yards. The juvenile whose querulous notes had previously attracted attention was now released from the shell; no change, however, could be detected in the remaining egg. Having now secured as many specimens as were needed, this egg was left, and though the old birds were seen an hour later some distance from the nest, the young one was eventually brought out."

Being frequently watched in order to observe the various changes undergone by juveniles of this

* Grebe are known to the Norfolk marshmen by this title.

species, the small family grew far less timid, taking in course of time but little notice of the punt at the distance of forty or fifty yards. This brood in their infancy corresponded precisely with the young figured by Gould, with the exception that the bare patch on the crown of the head exhibited a pale flesh tint, instead of the deep red depicted in the plate, that hue not being assumed till three or four hours after death. Later in the season I captured another young bird, out of a brood of four, probably ten days or a fortnight old, and endeavoured to rear it, small fish and water-insects being administered as food. Nothing that we offered appeared palatable; and after surviving two or three days without nourishment, the unfortunate became gradually weaker, and in the end succumbed to exhaustion.

A few lines extracted from my notes will give an idea of the food on which this species subsists while resorting to the freshwater broads:—"July 5, 1873. While fishing on Ormsby Broad, I noticed a number of broods, some newly hatched, while others were almost full-grown. Whether large or small, these juveniles appeared equally helpless, making not the slightest attempt to capture fish for themselves, but depending entirely on their parents. The youngsters followed at a respectful distance, the old birds diving occasionally, and when reappearing with prey one or other of the family would instantly swim up and claim its share. A keeper in the district, considered an authority on such matters, assured me that Grebes always rear their young on eels; to-day, however, I watched numerous broods fed repeatedly, and the food supplied appeared in most instances to consist of small roach. The operation was watched through powerful glasses at the distance of from forty to sixty yards, so there could scarcely have been any error as to the identity of the captives, a clear view being obtained of the fish while in the beak of the captor.

"July 25. Wind south-west, still and fine; the surface of the broad like a sheet of glass at daybreak. A pair of old Grebes (now showing a slight diminution in the frill) were seen with two young; as the latter were in a state (apparently between three and five weeks old) not previously examined, I secured both with but little difficulty, when the difference in size proved remarkable, one being double the weight of the other. Both exhibited a patch of bare flesh on the crown of the head, as well as in a line between the eye and the gape. The beak was black at the base on both mandibles, the remainder white with a narrow black mark near the point. Iris grey; legs and feet yellow and black mottled. Breast feathered; head, neck, and back still covered with down, no feathers showing on the wing. The stomach of the larger only contained a few feathers, apparently from the old bird, while the smaller had in addition a couple of perch between two and three inches in length."

The courtship of this species is an exceedingly amusing scene, and is referred to repeatedly in my notes. Under heading of March 6th, 1873, I find the following:—"This was the first day of the season that Loons were noticed on the broad, a pair in the Hickling corner proving unusually fearless. For over an hour I watched them sailing round one another, and occasionally pulling up and bowing their heads in the most singular manner. Whenever a halt was made the two birds swam up alongside and at once brought up, so that each faced the opposite direction. The grunts and squeals they gave vent to and the antics gone through need to be heard and seen to form the slightest idea as to their quaintness; to describe them accurately is utterly impossible." On May 15th, 1883, weather fine and hot, I again witnessed a most singular demonstration. The birds were evidently engaged in courting, though from time to time they desisted, and proceeded to wash and clean their plumage, the movements the pair went through being almost precisely identical. To conclude the performance, each trimmed the feathers of the other, like the Guillemots and Gannets repeatedly watched on the Bass Rock; I remarked that the bird operated upon stretched its head upwards while the other paid particular attention to preening the short feathers of the throat and neck. These noisy greetings not unfrequently take place in the reed-bushes; I have repeatedly listened to pairs calling loudly and to the splashing of the water, though the birds themselves were invisible. From the beginning of March till the middle of May the discordant sounds uttered by the Grebes may be heard in still weather.

Having occasionally fallen in with these birds in early spring exhibiting a curious mottled state of plumage about the head and neck, the colour of the iris also being of various tints, from lemon-yellow to orange, I am inclined to believe that the young do not assume the full crest till after the age of one year. As none but birds in full summer plumage are seen about their summer-haunts till the downy youngsters make their appearance, it is probable that, after the fashion of many of the sea-fowl, the immature birds are not allowed to associate with the adults, and necessarily seek other quarters. On the 11th of March, 1871, a small party of half a dozen were observed on Heigham Sounds: the birds proving shy, it was impossible to obtain a satisfactory view through the glasses; and being anxious to ascertain their state of plumage, I fired the big gun and secured four. Though the specimens varied considerably it was improbable that any would have assumed the full adult dress that season. On one occasion so early as the 18th of January, and often in February, I met with the Great Crested Grebe in full summer plumage. Doubtless the frills increase in size and richness of colouring with the age of the birds; I remarked that the elongated feathers of the ears as well as the crest of a pair shot in the east of Norfolk on the 25th of May, 1870, were little more than half the length of those on an old male obtained near the same spot on the 30th of May, 1873. I find only one entry in my journals referring to the date at which the adults commence to throw off the crest and assume the winter plumage:—"September 25th, 1879. The Loons had now lost the greater part of the frill and could hardly be distinguished from the young at the distance of eighty or one hundred yards."

By closely watching a nest discovered on Heigham Sounds in May 1870, I ascertained that the eggs were laid on alternate days; subsequent observations also tended to prove that this is the rule. When fresh laid, the egg is a pale bluish white, though it speedily assumes a dirty yellow tint, stained by the decaying water-plants with which the bird artfully conceals its treasures when leaving the nest.

As previously stated, it was seldom during the winter months that I observed these birds frequenting their summer-haunts; one or two, however, occasionally return, and having alighted on the ice, utterly bewildered by the change, flutter and slide along the surface from which the snow has drifted, and experience much difficulty in rising again on wing. The Plate represents a scene of this description witnessed on Hickling Broad in December 1871; the specimens from which the figures are taken were obtained at sea, off Shoreham, while in pursuit of small fry in the shallow water over the sand-banks on the 9th of December, 1879. The following were the colours of the soft parts:—Iris bright cerise, with a white ring round pupil. Upper mandible pale flesh, with a dark line along the ridge. Legs dark greyish green on the outer, yellow on the inner side. Toes yellow with a green tinge, showing clouded blotches and a dark scrawl down each toe. Nail on each toe a pale lead tint.

RED-NECKED GREBE.

PODICEPS RUBRICOLLIS.

THERE is, I believe, no recorded instance of this Grebe having been met with breeding within the limits of the British Islands, and I am unable to give any information that would assist in proving they ever remained as residents on the freshwater broads or the remote lochs of our northern or eastern counties during the summer months. That the juveniles make a start at an early date from the distant haunts in which they have been reared I have, however, good evidence, a young bird in the immature plumage of the first autumn having been obtained on Breydon Water, near Yarmouth, on the 11th of August, 1873. The adults also quit their northern summer-quarters at much the same time. On the 3rd of August 1872, I obtained a good view of a handsome old bird in full plumage about a mile off Brighton; it had escaped observation owing to the swell, and did not rise from the water till the lugger, in which we were proceeding to sea on a conger-fishing expedition, was within the distance of ten yards; having, however, unfortunately omitted to charge my gun, the chance of a shot was lost. After having been marked down about a quarter of a mile to windward, we failed to catch a glimpse of it again, the breeze having freshened and heavy seas occasionally rolling in and breaking.

Though luck had turned against us while in pursuit of the Grebe, we met with very great success when the marks * over the wreck, about which the big eels resorted, were reached. It had been reported a few days previously by one of the Brighton boatmen that a conger weighing seventy or eighty pounds had broken away from his line at this spot, after being struck by the gaff, which had left a deep and conspicuous wound in its neck. The whiting proved exceedingly numerous and kept us constantly at work, while they swam round and round in shoals about the blocks of stone that had been the cargo of the vessel and now afforded shelter for the congers to repose in. For an hour or so our baits were only taken by congers of twenty-five or thirty pounds, but at last a line was seized by one that strongly resisted all efforts to bring him to the surface. At length after a delay of a quarter of an hour, during which time he probably held on by one of the blocks of stone or pieces of old timber, he came in view, and the cut below the gills was plainly seen as he rolled round. After darting down again two or three times with irresistible force, he floated quietly up, apparently beginning to feel the effects of his struggle to escape, and was safely lifted into the boat by three of our crew armed with gaffs and well acquainted with such work. Although six feet one and a half inches in length, his weight was ten or twelve pounds less than what we had been led to expect; the fine condition, however, in which he proved to be accounted for his strength and the protracted resistance offered. Our total score that day amounted

* As it may not be generally known to those unacquainted with sea-fishing, it is well to state that the "marks" referred to are usually conspicuous objects on the shore, such as church-steeples, lofty chimneys, or high buildings, as well as hills and even tall trees that can be brought in line on both quarters, so as to assist the fishermen in finding the whereabouts of the rocks, wrecks, or sand-banks that prove so attractive to the finny tribe.

to six congers, the other five weighing between twenty and thirty pounds, with five-and-twenty score of rock-whiting and five score of other fish.

The summer plumage appears to be retained later in the season by this species than by its relative the Great Crested Grebe. While brought up fishing on the Church rocks off Shoreham, on the 13th of September, 1881, I noticed an adult fly past our boat within half a gun-shot, but being engaged at the moment in hauling up the line, I was unable to snatch up the breech-loader in time to secure it; the bird appeared exceedingly brightly coloured, none of the tints assumed during the breeding-season having yet faded. Again, about a fortnight later, on the 4th of October, while brought up at the same spot, another passed, making its way west, in almost similar plumage, though slight signs of a change in the bright red tints of the neck seemed to have taken place.

I have met with but few of this species during the winter months, but succeeded in shooting a fine specimen, exhibiting the plumage put on at that season, off Shoreham, on the 28th of January, 1881. This was a few days after the terrible gale that commenced on the 18th of the month and was followed by the excessive cold that caused such destruction to all small birds; the continued severity also drove thousands of Geese and other Wildfowl into the Channel along the south coast, and many Grebes of various species were observed at the same time within a short distance of the shore. On the 10th of December, 1870, the weather, though exceedingly cold with a heavy frost, being fine with scarcely a breath of wind, I met with an unusual number of Grebes collected at low-water, diving for food over the sandy flats between Lancing and Worthing. A few of different species were obtained as specimens, among which was a Red-necked Grebe; judging from its small size, more dingy colouring, and the absence of gloss on the feathers of the breast so conspicuous on some adults, I am inclined to believe it was an immature bird in the plumage of the first winter.

The juvenile which, as previously stated, was obtained on Breydon on the 11th of August, 1873, proved exceedingly wary, and a whole day was passed in the attempt to make a successful shot, and even when fatally wounded he escaped for a time. The diminutive stranger was first sighted on the deep water at the west end of the flats, diving among the masses of dense green weed that form such a favourite resort for grey mullet; on the approach of the punt it at once ducked below the surface, and in some manner succeeded in concealing itself in the tangle. Though three boats with my puntmen, all experienced gunners and well up to the work of searching for crippled fowl, surrounded the spot at the distance of forty or fifty yards apart, no signs of a ripple on the surface or even the slightest movement among the weeds could be detected during the hour and half we remained on the watch. As a last resource we moved to the bank on the north side of the flats, and, by the help of the glasses, kept the surface of the water at the spot we had lost sight of the bird under constant supervision for about two hours without any result. Then, having produced refreshments, another hour was passed; on looking again with the glasses soon after our repast was finished, the small diver was discovered swimming slowly about here and there on the open water among the weeds, and occasionally plunging below the surface, evidently busily engaged in pursuit of food. The punt was next sculled silently towards the spot, and without having raised the slightest alarm, a shot was fired from the big gun at about the distance of fifty yards, and the smoke clearing off, the bird was lying apparently helpless on the water. Great was the astonishment when, on an attempt being made to pick up our prize, it gave a short struggle and at once disappeared below the surface, and remained invisible during the whole of the afternoon, not a sign of its presence being detected, though we waited till darkness had set in. Having determined to rake up and thoroughly search through the weeds * about this part of the water, we made an early start the following morning. On arriving at the spot, however, soon after daybreak, the bird was discovered floating dead in a small open pool among

* On a former occasion a wounded Mallard, that dived and got suspended in the weeds in this part of the water, was secured by these means.

the weeds within a yard or two of where it had been lost sight of. There is little doubt that it had got entangled in this dense mass of twining water-plants, and, unable to free itself owing to the wounds inflicted by the shot, had finally been drowned. This specimen proved exceedingly small, apparently not half the size of the bird figured in winter plumage; the pinion-feathers of the wing also were very short, perhaps scarcely full-grown—so powerless, in fact, that it seemed a mystery how the juvenile had succeeded in making its way across the stormy seas from the northern lakes where the species is reported to breed. I did not take the weights of either the large or small bird, but the length of both is recorded and also the measurements of the wings, which may perhaps be of service in giving some idea as to the size of the two specimens.

The immature bird on Plate I. was fifteen and a half inches in length, from the carpal joint to the end of the primaries five and a half inches, second joint of wing three and a quarter inches. The bird in winter plumage on Plate II. was twenty and a half inches in length, from the carpal joint to the end of the primaries seven and a quarter inches, the second joint of wing four and a half inches. The upper and lower mandibles of the immature bird were both of a brown-horn tint, which was lighter towards the point, and with bright yellow around the base; the iris pale yellow; the bare line between the eye and the base of the beak a brown flesh-tint. Legs a greyish-green tinge on the outside, and a paler tint of the same colour on the inner side; webs clouded with yellowish spots or blotches.

The Red-necked Grebe represented on Plate II., showing winter plumage and obtained off Shoreham after the gales of January 1881, was easily secured, but a very short time being spent in his pursuit. The bird was sighted swimming by itself about a quarter of a mile at sea, late in the afternoon, just as our punt had been turned to make for the shore. Paying little or no attention to the approach of our craft, the shoulder-gun was brought into use, as the heavy swell rolling towards the beach would have rendered the aim of the punt-gun very uncertain. Luckily the bird was turned over perfectly dead without a struggle, as had it dived there would probably have been no little difficulty in securing it in the broken water. This was a fine large bird, nearly as weighty as many of the Great Crested Grebes I have met with in winter. The mandibles, which were strong and sharply pointed, were a pale brown or horn tint, lighter towards the point, the upper exhibiting a black line for about half its length, the base of both bright yellow. Iris a very pale straw- or lemon-yellow tint, streaked with lines like crystal, which rendered it almost colourless; the base-line between the eye and the base of the beak a deep red flesh-tint. Legs a dark brownish green on the outside, a lighter shade on the inner side, and the webs the same colours with pale yellow blotches diffused over them.

I have met with but few chances of making observations on the colouring of the eyes of this species in summer plumage, but one fresh-killed bird having been examined. In this individual the iris was perfectly colourless, a pearly white with lines like crystal radiating from the pupil and giving a very brilliant appearance to the eye.

The description of the colouring of the iris of the Red-necked Grebe by several writers is somewhat conflicting and decidedly at variance with my own experience. In Gould's 'Birds of Great Britain' we are told, where reference is made to the adults, that the "irides are red;" and the same author states, "The young bird of the year has neither the red neck nor the elongated head-feathers; the throat is brownish; irides brown." This is all at variance with the notes I made on the juvenile obtained on Breydon in August 1873 and figured on Plate I. "The irides red" is also the description given of the eyes of the adults in the fourth edition of Yarrell, and a still more recently published work refers to them as "irides brownish red."

Mr. Gunn, of Norwich, who is well acquainted with this species, has just sent me word that on the 22nd of October, 1885, he received an adult female of the Red-necked Grebe, shot the previous day, and the eyes proved to be white. Over twenty years ago Mr. Gunn called attention to the colour of the eyes of this species in his notes to the 'Naturalist,' vol. iii. page 30, published at Huddersfield in February 1865, stating that the eyes of an adult killed in February were pearly white, and also referred to the fact that the eyes of the young birds which had come under his notice were invariably a pale straw-colour. He adds that during the past twenty years over thirty examples of this species have passed through his hands, and, after having made inquiries concerning other specimens, he believes it improbable that a bird has ever been obtained in Great Britain with either red or brown eyes. In conclusion, I can only state that Mr. Gunn's experiences concerning the irides of the Red-necked Grebes, both adults and immature, agree in every point with my own, and there is no denying the fact that the specimens we examined have in all cases been fresh killed.

SCLAVONIAN GREBE.

PODICEPS CORNUTUS.

While collecting in the Highlands in the latter end of May 1869, I received word from a keeper that a Grebe with which he was unacquainted had taken up its quarters at a small loch near the west coast of Ross-shire. His description first led me to believe it was a Red-necked Grebe; and a drawing of that species having been shown to him, he stated that it appeared to be that bird. After a second and nearer inspection through the glass obtained on his next visit to the loch, the remarks made by the man convinced me that if a Grebe at all and not a Diving-Duck, it must be the Sclavonian. Being fully engaged with other matters, I had not sufficient time at my disposal to look thoroughly after this interesting stranger, and consequently the keeper was despatched to watch the spot closely and ascertain if a pair giving indications of remaining to nest in the district could be detected. One bird only was observed, though this was almost constantly in view while he remained on the look-out; when at length I obtained a chance to visit this lonely and wind-swept loch, some laddies tending the farmers' cattle had brought down their beasts to the waterside, and the object of our search (of which only a momentary glimpse had been caught) was driven to seek the shelter of the stunted bushes and beds of rushes that fringed the shore. Another attempt was made later in the day; and on this occasion blinding squalls, with drifting sleet and rain, followed one another in rapid succession; and in the end we were forced to leave the spot without having satisfactorily identified the bird. Rough weather had evidently set in; and the westerly swells rolling across the open water towards the only part of the loch where sufficient shelter for a nest could be obtained, rendered it utterly useless to prepare the india-rubber boat to explore the cover, as, owing to the foam lashed up by the squalls, it would be impossible to catch sight of the bird. From information received to-day (March 10, 1885) from the keeper previously referred to, I learn that the Grebes, which still remain unidentified, annually frequent the same piece of water (Loch-na-Feadernach), and have regularly reared their young. Last summer two pairs were seen for the first time on the loch, and later on a couple of broods appeared. I also heard from the owner of the ground that another pair of Grebes, evidently of the same species, frequent a small loch two miles further to the north, and breed there every year. As I have received word that if a specimen can be obtained it shall be forwarded to me for identification, there is hope that, if I am not able to revisit the locality, the uncertainty as to the species may still be solved. These are the only instances that have come to my knowledge where the Sclavonian Grebe might reasonably be imagined to have bred in this country. In June 1870 I noticed another in full plumage on one of the Norfolk broads; but on making inquiries it was ascertained that during the previous winter the keeper had knocked down a bird of this species, and discovering shortly after that his prize had recovered from the effects of the shot, it was pinioned and then turned out on the broad.

This species has come under my observation during one season or another in almost every county I have visited. Along the Sussex coast specimens are to be seen both in spring and winter, and the same remarks also apply to Norfolk and Suffolk. In the east of Ross-shire and also in Sutherland, on both the fresh- and saltwater lochs and firths, I met with these birds repeatedly in spring, but never noticed a single specimen during autumn or winter. The only occasion on which this Grebe was supposed to have been identified in the Western Highlands has already been referred to; other diving-birds somewhat resembling these Grebes have been reported, but those from whom my information was drawn could scarcely be trusted, their descriptions of plumage and habits being vague and in some instances contradictory. An exceedingly fine male in my collection was obtained on Loch Slyn, in the east of Ross-shire, on the 26th of March, 1869. A strong breeze from the east was blowing when a pair of Grebes were sighted within thirty yards of the bank; one, I ascertained by the aid of the glasses, was in the full adult dress, while its mate (evidently a female by the size) exhibited not a single stain of colour on a plumage of spotless white. Not having any desire to add such varieties to my collection, I secured the brightly coloured male with the first barrel, when the albino vanished from sight, and the swell rolling across the loch put an effectual stop to any chance of again detecting the stranger. During the next ten days, on the Dornoch Firth and also at the Little Ferry, I met with several fine and highly coloured birds; with the exception of a female needed as a specimen, these were allowed to escape; in every instance they proved utterly unsuspicious of danger. The smaller Grebes are almost invariably fearless till molested; if once alarmed it will, however, be found by no means easy to obtain a shot. For safety they trust almost entirely to diving, their actions below the surface being exceedingly puzzling; a wounded bird may be watched in clear water turning and twisting with the greatest ease, its singular lobes or toes being used as paddles and feathered between each stroke in the most rapid manner.

A severe winter often brings many of these birds in the plain and unpretending dress assumed at that season along the east coast, several finding their way on to the broads and rivers when free from ice. The flat sandy shores of the numerous marshes along the coasts of Kent, Sussex, and Hampshire are frequented at the seasons of migration by numbers of Grebes of all species, the large extent of shoal water at low tide enabling them to secure an abundant supply of small fish and marine insects, Rye Bay and the mussel-banks off Lancing being exceedingly favoured resorts. At times I have noticed small parties of these Grebes, for the most part young birds, in the pools of brackish water inside the shingle-banks and also in the sluices and marsh-dykes.

My notes contain descriptions of many changes of plumage exhibited by this species, as well as the colours of the soft parts at various seasons: to give all these is scarcely necessary; it may not, however, be out of place to state that the line of bare flesh which extends from the eye, and runs with a slight break at the gape about half down the lower mandible, is conspicuous at all seasons and is one of the means by which this species may be distinguished from the Eared Grebe. In the full plumage this flesh-mark is of a deep reddish tint, though in winter it fades considerably in colour.

EARED GREBE.

PODICEPS NIGRICOLLIS.

Though possibly not an unfrequent visitor during the breeding-season, in days gone by, to several of the broads in the eastern counties, this species has never come under my observation alive in summer. That they have nested and hatched their young in this country there is, however, not the slightest doubt, a full-plumaged adult and a couple of downy mites having been brought, as a great prize, by a marshman whom I had requested merely to look out for and ascertain the usual haunts of a pair that were reported to frequent a certain corner on a large sheet of water. The manner in which these specimens had been procured rendered them useless for my collection; and under the impression that others would be met with on some future occasion, I took no notes of the colouring or the markings on the soft parts of the juveniles: since that date, however, now nearly twenty years ago, I have only met with these Grebes after the unpretending dress put on at the close of autumn had been assumed.

In winter this species is occasionally to be seen off the Sussex coast as well as on the pools and rivers in the salt marshes. Some years back while staying at an inn in Pevensey Level, so as to be on the spot for Snipe-shooting, I learned that a small diving-fowl had been caught by a dog; and on visiting the back kitchen, where the captive was disporting itself in a large tub of water covered with a net, I found what appeared to be a winter-plumaged Sclavonian Grebe *. In order to make observations on its powers of locomotion on dry land, the bird was fished out and set at liberty on the brick floor: while shuffling round with the wings spread I succeeded in taking two or three rough sketches; these drawings now plainly indicate the species, viz. the Eared Grebe. The white bar on the wing of the Sclavonian Grebe is merely a patch of about an inch and a half in length, while in the present species it extends over two joints: this very conspicuous mark renders it possible to distinguish the birds, even in winter plumage, while on wing. On the 10th of December, 1879 (a light breeze from the north, a sharp frost setting in after midday), I was shooting off the coast between Shoreham and Worthing, and at half-ebb came in among the old groins and breakwaters which stretch out to sea for a considerable distance. An attractive harbour for small fish, shrimps, prawns, and other marine insects is here afforded by the shelter of the decaying piles and planks that still hold together, and food being always abundant in fine weather, numbers of Grebes and Divers are drawn towards this happy hunting-ground. During the course of the afternoon I plainly identified every species of our British Grebes, obtaining specimens in full winter plumage of the Great Crested, Red-necked, and Eared, as well as passing and closely examining several Sclavonian in the shoal water near the sands. On landing and mounting to the top of the beach, to obtain a better view of the bay through the glasses, a couple of Little Grebes were observed busily fishing in a brackish pool beyond the shingle-banks. One or two other specimens of the Eared Grebe shot by local gunners along the Sussex coast have come under my observation during the past few years: these all exhibited the full winter plumage.

* I had never previously met with this Grebe in winter, and was unacquainted with the characteristics that mark the species.

Though the Norfolk broads are generally supposed to have been the stronghold of this Grebe in the British Islands, I have not observed any of these birds about the saltwater estuaries or rivers during winter, and but a single specimen was met with during the years I passed on the east coast at that season. Shortly before daybreak on the 9th of November, 1879, while steaming through the St. Nicholas gat direct for the open sea, a bird was distinctly heard to strike once or twice immediately above our heads against the iron railings on the bridge and fall apparently on deck. Though no signs of the unfortunate could be discovered at the time, an Eared Grebe was found in the small boat when lowered to pick up some wounded Skuas. The head, which had evidently been trampled upon by one of the crew, appeared to have suffered from previous injuries; there was, however, no difficulty in ascertaining the species, as the wings were still perfect.

As the winter plumage of the Eared closely resembles that of the Sclavonian Grebe, the following remarks may be of service in assisting to identify specimens obtained at that season. The upper and lower mandibles of the Eared Grebe turn slightly upwards towards the point, while the beak of the Sclavonian is perfectly straight; the latter also exhibits at all seasons a line of flesh from the eye towards the gape, which is not apparent in this species. As previously stated, the white patch on the wing extends over the upper and lower joints in the Eared Grebe (A), though in the Sclavonian (B) it is merely visible on the upper; this distinction can be easily recognized in the woodcuts.

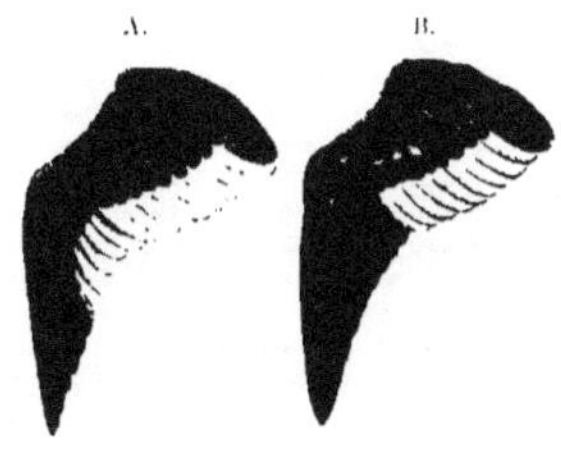

LITTLE GREBE.

PODICEPS MINOR.

THIS diminutive Diver is widely distributed over the British Islands; I met with numbers on the shallow and reedy lochs in Caithness, Sutherland, and Ross-shire; several pairs also have been observed nesting about the rush-grown pools and ponds in the east of Sussex. It is somewhat strange that in the east of Norfolk, where the larger broads might naturally be supposed to offer attractions to this species, the Little Grebe is by no means abundant, the few that I have met with being observed for the most part about the dykes and small rush-grown pools in the marshes. As their haunts are generally surrounded by low cover, such as elder, hazel, willow, or brambles, below which they seek shelter when alarmed, it is probable that, though the fact of their residence in these parts usually escapes the observation of the natives, a few pairs rear their young in the district. In Sussex I remarked of late years that this species appears to have considerably decreased in numbers, many of the pieces of water to which these birds were regular summer visitors being now entirely deserted.

Though difficult to force on wing, an adult Little Grebe can fly strongly and at a rapid pace when once a start has been effected; but few opportunities for watching them making their way from one spot to another have, however, fallen to my share. A pair that nested on a shallow and rush-grown pond of about a quarter of an acre near the church in the centre of the village at Falmer, near Brighton, regularly made their appearance on the approach of summer: their arrival and departure probably took place during the hours of darkness, as I could gain no information concerning their movements from even the workmen constantly employed near at hand.

In autumn a partial migration towards the south undoubtedly takes place; at this season small parties numbering from three or four up to eight or nine, or even a dozen, have occasionally come under my notice within a short distance of the shores of the Channel. The river and pools in the marsh between St. Leonards and Bulverhithe, the backwater at the sluice in Pevensey Level, and the flooded meadows between Lancing and Shoreham are the spots where the largest gatherings have been witnessed. On one occasion a party of ten were seen swimming and playing on the open salt water just off the breakwaters to the west of Shoreham. Towards the latter end of September 1878 Little Grebes were unusually abundant along the south coast; for several consecutive days I noticed at least a dozen on a piece of brackish water that had collected inside the shingle-banks. When threatened by danger the birds dived in towards an embankment thrown up to keep back the encroaching tide, and concealed themselves among some faggots staked down to assist in resisting the wash of the water. In several instances, on approaching the spot near which one had been lost sight of, their bright eyes were detected below the surface, their bodies being almost entirely hidden among the weeds and rubbish gathered between the faggots. On retiring quietly from the vicinity of their place of concealment, but few minutes elapsed before they were again busily diving for food, black bugs and various other insects of repulsive aspect being probably the attraction that drew them to the spot*. Four or five young birds were taken about the same time by a shrimper who was catching "hunters"† up one of the dykes running through the marshes towards Shoreham harbour.

I am aware it has been stated by some authors that the Little Grebe can walk with ease, standing at the same time upright and proceeding at a moderate pace. Though several have at times come under my observation squatting among the bright green grass on the spits of land that extend into some of the northern lochs, I never noticed the birds at a greater distance than three or four feet from the water. On leaving or returning to their natural element, whether alarmed or of their own accord, they invariably appeared to shuffle over the ground without raising the breast above an inch or two from the surface. So plentiful were these birds some years back in the east of Ross-shire that on one occasion, in April 1869, I counted three pairs sitting among the herbage within the distance of sixty or seventy yards from one another. A couple that happened to be within range and were needed as specimens owed their lives to the fact that a Mute Swan was sleeping quietly on the bank less than a yard beyond them.

I repeatedly came across the nests of this species among the dense cover at a small loch in Cromarty with a Gaelic name I could never catch—a reed-grown stretch of mud and water a short distance to the south-east of Loch Slyn. These birds apparently commenced breeding early in April, the full complement of eggs being usually laid before the middle of the month. On the 27th of May, 1869, having ascertained by previous inspection that two or three broods had been hatched out, I was enabled by the help of an india-rubber boat to thoroughly explore the reeds and stagnant pools as well as the whole of the open water. For several hours no success was met with, and at length, utterly worn out by the labour of struggling through these almost impassable swamps, I made myself as comfortable in the buoyant craft as circumstances would permit, and sought rest and refreshment while watching the actions of a pair of old birds that for some time had closely followed my movements. After waiting for over an hour and finding that the Grebes evinced no signs of taking their departure, I worked the boat suddenly in their direction, when a single downy youngster was detected and speedily bagged by a small breech-loader. On turning to look round the pool on which I had been brought up, two more were at once discerned floating perfectly dead on the water. These must have dived below the surface during their efforts to escape, and having risen underneath the boat would naturally perish by drowning, not possessing sufficient strength to free themselves from the inflated cylinders surrounding the bottom. The tiny mites had evidently but recently left the egg, their beaks exhibiting a fresh rose tint tipped with a paler hue, the legs and feet being of a blue livid flesh.

On this loch the whole of the nests I met with were floating on the surface, moored among the reeds, rushes, or tangled water-plants on the deeper and more open parts of the pool. At the larger sheet of water known as Loch Slyn, and distant only about half a mile, all I examined were connected with the rank grass and other herbage springing from the banks.

* A specimen I secured contained at least a handful of these horrible creatures.

† Local name for a species of prawn that frequents brackish water.

GREAT NORTHERN DIVER.

COLYMBUS GLACIALIS.

Whether it is admissible to state that the Great Northern Diver breeds regularly within the limits of the British Islands appears to have puzzled many ornithological writers. From personal observations I cannot give any opinion on the subject, having as yet been unable to devote sufficient time to thoroughly explore the remote and lonely districts in which this interesting bird is supposed to take up its summer-quarters.

Adults and immature, both in various stages of plumage, are to be met with at almost all seasons on the saltwater seas, lochs, and firths surrounding our islands. During early spring, autumn, and winter many penetrate to the inland Highland lochs, firths, and rivers, as well as to the lakes, broads, and estuaries of the English counties. When unmolested, I frequently remarked that the juveniles were by no means suspicious of danger, and paid little or no attention to those who merely watched their movements. The old birds, however, are, in most localities, exceedingly wary, and, unless among the rocky islands off the northern and western coasts of the Highlands, where this species is seldom interfered with, offer few chances for observers to study their habits, or to collectors to secure specimens.

In a most unexpected manner, a few years ago, I met with an excellent opportunity for examining an adult in magnificent plumage, at close quarters, off the north-east coast of Scotland. During my residence at Tain, in Ross-shire, in the spring and summer of 1860, I frequently explored the Cromarty rocks, to take notes on the Peregrines, Herons, Rock-Pigeons, and other birds resorting to this wild and rugged range of cliffs. Several cavities, just above high-water mark, are to be found in the face of the rocks near Shandwick, and having made myself comfortable in one of these recesses, I was engaged in taking lunch in company with a couple of keepers. The day was fine, with scarcely a breath of air to ruffle the waves rolling up the cliffs, when suddenly, within five or six feet of the small ledge on which we had taken up our station, a Great Northern Diver skimmed up from the depths below to the surface of the water. The bird exhibited no signs of alarm, and as we all kept perfectly motionless, it remained for several seconds intently gazing in our direction, and then turning slowly and silently round, by the action of one foot, with the web expanded, the other foot being backed, dipped quietly down and continued its course for about eighty or one hundred yards before reappearing on the surface. The colouring of the eyes of this fine specimen was remarkably brilliant, the iris being of a bright crimson, sparkling with lighter rays of the same conspicuous tint around the pupil.

On referring to my notes for 1872, while shooting on Hickling Broad and Heigham Sounds, in the east of Norfolk, during November and December, I find these birds were seen repeatedly on those waters during the winter, specimens in immature plumage having been obtained for examination on the 28th of November and on the 4th and 14th of the following month. I only met with single birds, and all that came in view exhibited the earlier stages of plumage assumed during immaturity. But few words are necessary concerning the two first, as both were secured without difficulty; the third, however, proved an exceedingly tough customer, and after surviving charges from both punt- and shoulder-guns, endeavoured to avenge his injuries by an attack at close quarters. A shot with the punt-gun at once settled the specimen obtained on the Broad in November, and a green cartridge from a muzzle-loading 10-bore brought down the second, whose state I could scarcely judge at the first glance, while crossing the sounds on wing at a considerable elevation early in December. Our encounter with the third may be described by the following extract from my notes:—"December 14. All signs of frost vanished. On reaching the Broad, the surface of the water was as smooth as glass, and soon after daybreak we detected a Great Northern Diver on the deep water in the channel near Pleasure Hills, and, as the light increased, ascertained that the stranger was busily engaged in diving for food, though apparently meeting with but little success, as it was not until we were within seventy or eighty yards that the bird came up with a small pike of about twelve or fourteen inches in length. Here we stopped the punt to watch its actions, and remarked that the prey was grasped crossways in the beak and assiduously shaken for over five minutes, this treatment being doubtless administered with the intention of disabling or killing the fish before it was swallowed. At length, as the bird appeared a finer specimen than any previously secured, I fired the big gun at the range of about eighty yards, when, after beating the water violently with its wings for a few seconds, it dived, speedily reappearing again on the surface, evidently hard struck and uttering the most mournful cries I ever heard proceeding from the throat of a bird. Desirous of putting an end to its sufferings as soon as possible, I picked up a heavy 10-bore breech-loader, and as the Diver continued swimming towards the boat, discharged one barrel at the distance of between forty and fifty yards. Though lying motionless on the surface of the water for a moment after receiving the charge, it again flapped towards the punt, which had now turned round, and having made its way so close that a second shot must have damaged it as a specimen, the bird was allowed to continue its course, till, reaching the stern of our craft it scrambled on to the after-deck and was shuffling open-mouthed over the wash-streak into the interior, when it was seized by the neck by the puntman, who soon put an end to its struggles. The Diver was remarkable for its size, and proved also to be in exceedingly good condition, weighing just over 9 lbs.; the two previously obtained had only turned the scales at 8 and 8½ lbs. The iris was a deep olive-brown, the mandibles white, with the exception of a dark mark down the ridge of the upper, extending almost to the point. Inside of mouth a dirty livid white tint; legs outside black, inside white, edges light grey; toes black; webs white, with veins of purple tinge showing very conspicuously down the centre."

During my wanderings through the Outer Islands of the Hebrides, I again often listened to the monotonous wailing notes of this species. An immature bird that was particularly noisy attracted our attention one morning in May 1877, on a saltwater loch in the Park district of the Island of Lewis, while we were obtaining a shot at an old female White-tailed Eagle. This Diver evidently possessed vocal powers of the highest order, for it treated us to a most discordant concert for over half an hour, plunging below the surface occasionally, and again on reappearing uttering its mournful cries.

Full-plumaged birds are to be seen every spring in the Channel off the coast of Sussex; they first put in an appearance about the middle of April, and are to be met with throughout the greater part of May. Some seasons they pass along in immense numbers, all making their way from west to east, bound towards their breeding-grounds in the far north; the wind doubtless accounts for the course followed; occasionally they shelter under our shores and at times are forced to seek smooth water across the Channel. Adults and those in various immature stages often travel in company. On the 21st of April, 1874, I noticed half a dozen fine mature birds intermixed with three or four that exhibited only a mottled or half-and-half state of plumage, about ten miles at sea off Brighton; a few days later some Worthing fishermen informed us that they had met with thirteen in one party off Goring about the same date, half of which were in full plumage.

To obtain specimens of these Divers out on the open sea is by no means an easy matter, the birds invariably dive (or, it might be more correct to say, submerge themselves so rapidly, and the operation is performed with such little exertion, though they certainly go down head foremost, which has been denied) when approached within about one hundred yards, and though the surface of the water may be as smooth as glass, no signs of their whereabouts, not even the slightest ripple, can be detected till, when all is quiet and the boat moved off, they again show themselves at the distance of about half a mile. Occasionally I have noticed a few immature birds of this species during the winter months while gunning off the coasts of Kent, Sussex, and Hampshire; the last recorded in my notes is under the date of December 29th, 1883, when one was observed off Lancing, several Eiders and immense numbers of Scoters, both Velvet and Common, being on the water the same day.

While staying at Loch Inver, on the west of Sutherland, during the summer of 1877, I ascertained that Great Northern Divers in full plumage were often seen on many of the lochs along the coast, and that they evinced a strong partiality for Loch Roe, a small sheet of water shut in by high rocks and entered by a narrow channel, where, if necessary, a shot might almost with certainty be obtained at a bird attempting to escape towards the open sea. On the 7th of June, the weather being too rough to fish or put to sea in quest of specimens, I started with a gillie who was well acquainted with the haunts of the birds, to drive to the loch in a rickety old dogcart, designated a "machine" by the natives. Though I soon discovered our visit was made too late, the Divers having left for the season, and only a few Razorbills and one or two Red-throated Divers were seen on the water, I was amply repaid by the drive through such a wild and primitive country. Before reaching our destination we were obliged to leave the conveyance, as the rough track became too narrow for even that small and humble vehicle. The gillie assured me that the road, which was merely a slight excavation in the hill-side with a few of the larger blocks of stone rolled down, had been greatly improved upon during the past few months, and that previous to the alterations he himself considered it had been "very ill-looking." Neither windows nor chimneys, I remarked, were to be seen in the few shealings we passed, and the peat-smoke was forced to find an outlet by means of the cracks and crevices in the dry stone walls or through the doors*. The carts made use of by the natives were small and roughly put together, somewhat resembling the "trawl" and "barrow-carts" employed by fish-dealers and tradespeople at Yarmouth, which are fashioned to penetrate the narrow passages termed the "rows."

While engaged in obtaining a pair of Herons required as specimens along the face of the Cromarty rocks, in May 1869, one of them shot from the summit of a portion of the cliffs known as the Cairn Rhui, fell disabled towards the water, and dropped on the back of a Great Northern Diver swimming slowly towards the east, a short distance from the shore. The Diver evidently resented the injury, striking savagely two or three times at the cripple before ducking down below the surface; after a short absence he seemed disinclined to leave the spot entirely, reappearing again within the distance of sixty or seventy yards, and remaining on the watch till put on the alert by the approach of a salmon-coble sent out from the fishing-station at Shandwick to pick up the wounded Heron.

* Such dwellings are referred to under the heading of the White-tailed Eagle, p. 4.

BLACK-THROATED DIVER.

COLYMBUS ARCTICUS.

This handsome species is a summer resident on many of the larger sheets of water in the Northern Highlands, as well as a spring, autumn, and winter visitor to almost every portion of our southern and eastern coasts with which I am acquainted. In the choice of a situation for their breeding-quarters the Red- and Black-throated Divers differ considerably—the former delighting in the flat moors and floes to be found in Caithness and the east of Sutherland, while the latter prefers the grassy or sandy islands in hilly districts, or at least where the view is shut in by the lofty mountains whose rugged outlines prove an endless source of attraction to the visitors to those wild and romantic regions in the west of Ross-shire and Sutherland.

Without, I believe, a single exception, every nest (if such it may be termed) of this species that has come under my observation was placed on an island, coarse grass, moss, and heather usually forming the site on which the eggs were laid. The fact that these plants are broken down and killed by the weight and warmth of the body of the bird, and the depression gradually assumes the shape of her breast, has led to the idea that a nest is constructed. This is evidently a mistake, as in several instances, and more particularly on Loch Shin, I observed the eggs laid on the sandy shores of the small islands without the slightest attempt having been made by the birds to line their cradle, which was evidently only formed by their breasts while engaged in the labour of incubation. In one instance, on an islet near the centre of the loch, I remarked that the eggs were almost completely buried in the fine gravel which had been gradually worked over them by the body of the old bird while shuffling backwards and forwards from the water. This species, though usually wary and difficult to approach, becomes exceedingly confiding when unmolested, paying little or no attention to those intruding on their haunts *. I remarked this fact on several of the retired lochs in the deer-forests to which the tourist and the collector is seldom allowed to penetrate.

Even in summer adults in perfect plumage are often seen in company on the inland waters in the Highlands; on the 18th of June, 1868, half a dozen were observed on Loch Doula, near Lairg in Sutherland. These birds all proved exceedingly animated, chasing one another above and below the surface, and giving utterance while on wing or on the water to a variety of harsh cries and occasionally yelping like a dog. Diving and fluttering while pursuing or pursued, they repeatedly came up within thirty yards of where, in company with four or five keepers and gillies, we were sitting on the rough heather-bank by the loch-side. Some three weeks later the same summer, ten or a dozen, all in full adult plumage, were seen on Loch Craggie, a short distance to the east of the last mentioned loch. These birds were also sportively inclined, dashing about over the water with loud cries, till a party of eight or ten passing over, they rose

* The fearless behaviour of a pair watched for several hours on a loch in the west of Ross-shire in May 1868 is referred to under the heading of the "Common Gull," page 2.

and joined them, the whole making off in a straight line in the direction of Loch Shin. I remarked that Black-throated Divers, when flying in company, always kept in line one behind another at regular intervals. Whether these gatherings, which were occasionally composed of as many as fifteen or twenty birds, consisted of those that had been robbed of their eggs or young, or had merely met together for the sake of company, it was impossible to form an opinion, as possibly on my next visit to the loch on which they had been observed only a single pair would be visible.

The Black-throated Diver nests early in the season; I met with eggs almost on the point of hatching on a small island in a loch near the west coast of Ross-shire on the 21st of May, 1868. A second may possibly be sometimes reared; as late as July 7 in the same year I discovered two eggs on an island in Loch Craggie, a few miles to the east of Lairg. In all probability, however, in this instance, the bird had been robbed of her eggs once or twice earlier in the season.

On the Norfolk Broads these Divers are now and then seen in immature plumage or while undergoing the change into the adult dress, and the same remarks also apply to the species in the channel off the coast of Sussex. The old birds are seldom seen here in full plumage, though one specimen in a remarkably perfect state was observed off Shoreham in 1883, so late as the 6th of October.

Numbers of Divers, both Black- and Red-throated, were to be seen about the lochs in the Long Island at the time of my visit in May 1877, and both species were found to be especially abundant about the upper portion of Loch Seaforth. This fine sheet of water separates Harris from Lewis for many miles, and although the tide flows up to the head of the loch, a narrow passage, known as "the rapids" in the district, renders navigation impossible except at certain states of tide. The majority of the Divers appeared to shun these narrow falls: a fine adult Black-throated, however, made the attempt to pass up where the torrent was pouring down with tremendous force; again and again he faced the rushing waters and was carried back, but at last by sheltering under the stones and in the swirl of the backwaters by the banks he attained his object, and having surmounted all difficulties sailed majestically up towards the head of the loch.

The newly hatched young are covered with a thick dusky black down and, like all waterfowl, take to their natural element immediately after leaving the shell.

RED-THROATED DIVER.

COLYMBUS SEPTENTRIONALIS.

This is by far the most numerous of the Divers that frequent the seas around our shores or take up their summer quarters and rear their young in the more northern portions of the British Islands.

As a rule this species selects its breeding-haunts in flat districts, though I met with a few pairs on the lonely lochs in the rough hill-country to the north of Loch Maree in the west of Ross-shire. A pair were evidently breeding on Loch na Fad in May 1868; but though the water was closely watched for several hours neither eggs nor young could be detected, and a terrible deluge of rain with blusterous squalls of wind that burst upon us shortly after our arrival put a stop to all attempts to catch a glimpse of the birds on a second occasion, and rendered a speedy retreat from that dreary solitude at once necessary.

In June 1868, while procuring Divers as specimens for preserving, two eggs of this species were discovered on the edge of a piece of black peaty water near the Crask in Sutherland. Our inspection of their treasures appeared to have proved annoying to the birds, and a second pair were laid at the distance of about twenty yards from the first. These eggs were placed, as is usually the case with this species, within a few feet of the edge of the pool, the Red-throated for the most part choosing a cradle for its young on the mainland, while its Black-throated relative almost invariably selects an island.

The Red-throated Diver is especially numerous in the central and eastern parts of Sutherland and also on the flat moors of Caithness, breeding in great abundance about the floes in the centre of the county, the black pools of water and the swampy nature of the country occasionally rendering their eggs unapproachable. At the time of my visit to this locality, I found the portable india-rubber boat most useful in exploring these desolate swamps, and by its help I succeeded in reaching the haunts of every pair marked down on the open moors.

I have usually remarked that these birds are to be found in great numbers when sprats are plentiful off our coasts. On the 8th and 9th of January, 1880, the weather at the time being dull and cold, with a light easterly wind and sharp frost, the boats off Brighton and Shoreham, as well as along towards the west, off Lancing, Worthing, and Goring, were taking large hauls of fish. Red-throated Divers were exceedingly abundant on the water, and large flights were also passing towards the west, at least five hundred being seen on wing during the space of an hour at midday on the 9th, flying in flocks of from twenty to double that number and also in a continuous stream of single birds. Scarcely a winter has passed when I have been shooting in the Channel off the coasts of Kent or Sussex without large numbers coming under my observation previous to squally weather; they are then seen passing west in small parties, following one another in rapid succession.

The summer plumage is occasionally retained till late in the year; during the last week in October 1872, while steaming with the herring-fleet outside the Cross Sands off Yarmouth, I noticed some hundreds of

Red-throated Divers; many were in full, though the majority showed intermediate plumage, their necks being much speckled with white. The whole of the birds were collected within a short distance of the herring-boats, evidently attracted to the spot by the abundance of fish.

On the 24th of May, 1883, while at our boat-house on the shingle-banks between Shoreham and Lancing, in Sussex, I watched a Diver busily searching for food close to the shore. The bird was probably in quest of shrimps, and into such shallow water did his prey lead him, that having been caught by an unusually heavy roller while rising to the surface, he was turned right over and carried helpless on to the beach. In a moment, however, he recovered, and by the help of wings and legs, aided by a receding swell, was swept back to his natural element.

I repeatedly remarked during autumn and winter that immense flights of these Divers are observed in the Channel flying west shortly before stormy weather sets in; on such occasions Guillemots and Razorbills are also following the same course. With hardly a single exception, I find by referring to my notes that these general movements foretell a change, heavy gales usually setting in within a few days.

A somewhat ludicrous incident which might, however, have been attended with serious results, occurred while I was engaged in procuring one or two specimens of this species in winter plumage in the Channel early in 1870. A wounded bird which had led us within a short distance of the beach off Aldrington, a mile west of Brighton, was on the point of being secured, when our attention was attracted by the report of a gun or rifle on the shore, and the next moment a bullet flecked up the water twenty yards from the boat, and passing harmlessly by struck again about thirty yards beyond. The wind was north, and the sea inshore as smooth as glass, and on looking from one spot to the other I could not understand how our boat or crew escaped without injury. A dense fog obscured every object beyond a hundred yards, but a figure from which the shot had evidently proceeded was dimly visible on the beach; and as one good turn deserves another, I picked up a heavy double charged with 1¼ oz. green cartridge No. 1 and returned the compliment. The old gun had a nasty trick of balling, and one of the men remarked we should probably find some part of the delinquent on the shore. He was just vanishing in the haze as I fired, and when we landed no signs of him could be found. Some navvies at work at a road near at hand were attracted to the spot by the shots, and on an inquiry having been made informed us that they had seen a person armed with a long gun firing out to sea on several occasions during the morning. Induced by the offer of a reward the whole party started off in the hopes of securing this reckless individual; in this, however, they failed, though returning with a couple of gunners, whose unstained weapons and pouches of small shot proclaimed their innocence, and whose evidence confirmed that of the workmen as to the escape of the man previously alluded to. On our way back towards Brighton we happened to run alongside a shrimp-boat that had been trawling in the same bay, and one of our men replied in answer to their inquiry "What sport?" that we had had a narrow escape of being bagged ourselves. "What!" exclaimed the fisherman, "you have been shot at? Look at our sails—two bullets through the for'sail, two between the masts, and another under the counter. My mate says: 'Up trawl, Joe, and let's hook it. 'No,' says I; 'may as well be shot as starved. Let the beggar shoot.'" This doubtless was the work of the same "cuss," who, as a Yankee would express it, must have had "snakes in his boots" that day. To offer any explanation for such conduct is utterly beyond my power, though I have heard of similar cases of insane thoughtlessness occurring along other parts of our coasts: I can now only regret we failed in our attempt to secure the perpetrator.

COMMON GUILLEMOT.

URIA TROILE.

In former days this species was known in many parts of the country as the Foolish Guillemot, its confiding nature and utter disregard of danger having led to this uncomplimentary appellation—to imagine everyone a rogue till proved to be a fool (by no means an unsafe rule to follow) having evidently been the creed of those who bestowed the title. The birds that frequent many of the southern breeding-stations in the British Islands have greatly diminished in numbers of late years, owing to the persecution to which they have been exposed by thoughtless gunners, whose only object was useless slaughter. Among the islands off the Western Highlands, and also along the western and northern coasts of the mainland, they, however, still hold their own in colonies of countless thousands, and little fear need be entertained of any diminution taking place unless by disease.

During autumn, winter, and early spring I have often observed large flocks of Guillemots off different parts of our southern and eastern coasts. At times these bodies appear stationary for a day or two, moving backwards and forwards with the tide, or following the course taken by the shoals of herrings or sprats; as a rule, however, they generally make their way steadily in one direction. These movements are regulated by the season—south in autumn over the North Sea, and then passing on west through the Channel, the course is reversed in the spring. Many, however, must remain behind off the shores of the Highlands and still further north, as a day or two previous to stormy weather setting in during winter they become exceedingly restless, and large bodies of this species, together with Razorbills and Red-throated Divers, may be seen a few miles at sea, in the Channel, following one another in rapid succession, all moving towards the west.

On the 19th of August, 1874, while staying at Canty Bay, in East Lothian, I remarked that immense numbers of adult and immature birds were to be seen on the water in the firth, all the ledges on the Bass Rock being then deserted by them. This was, however, by no means an early start from their breeding-quarters, as I find in my notes for 1873, under date of August 9th, that the water outside the Cross Sands off Yarmouth, and again almost as far north as Cromer, was in some parts almost black with these birds and Razorbills. The whole of the sea-fowl at the time we passed them were evidently resting quietly on the waves; no general movement was undertaken on this occasion, the shoals of fish gradually working south being doubtless the attraction to this favourite feeding-ground.

On the 23rd of April, 1874, thousands of this species were observed in the Channel, some nine or ten miles out at sea, off Brighton. Several large flocks flew east during the day, but the majority were floating motionless on the glassy surface of the water, unruffled by a breath of air. These birds were, with but few exceptions, in full summer plumage, only a single specimen in the perfect winter dress being noticed.

Considering the size of the pinions, the speed at which this species makes its way when once on wing is surprising. To rise from the water is, however, a somewhat lengthened operation, the feet of the bird flapping along the surface for some distance before it is enabled to make a fair start. When approaching their breeding-quarters at any elevation on the cliffs, they are forced to make several extended circles on wing, gradually mounting each time till a sufficient elevation is attained to allow them to drop on their respective ledges. The circuit of the Bass is generally made a few times before they are enabled to reach their quarters, generally situated near the summit of the northern and western faces of the cliffs. The Pinnacles at the Fern Islands are a group of small square crags of rock, of no great elevation, to which hundreds of pairs of Guillemots resort for breeding-purposes during summer; the flat tops of each are almost entirely covered with eggs, and the birds usually succeed in reaching the highest points by flying straight to the spot. When leaving their ledges at the Bass, I remarked that these birds dropped a considerable height before appearing to gain any assistance from their wings and launching out towards the open sea. When the dense ranks assembled on the Pinnacles are alarmed by the approach of a boat or a shot, a scene of indescribable confusion ensues: a general and hurried start is made, some take wing at once, while others either dash down or fall helplessly into the water, their flight being impeded by the swarms pressing on and endeavouring to follow in the same direction; eggs and young are occasionally sacrificed, and being rolled from the rock, drop down to the water, suffering not unfrequently from contact with the projecting ledges in their descent.

While inspecting the various species of sea-fowl breeding in countless thousands along the face of the cliffs between Dunnet Head and Duncansby, in June 1869, I noticed that the lowest nests (those at an elevation of only thirty or forty feet above the waves) belonged to the Kittiwakes, their cradles being clustered thickly together, ten or a dozen up to even a score forming a single mass of seaweed, in most cases well suffused with guano. Razorbills occupied the highest stations immediately below the summit, and Guillemots were scattered here and there over the central portion of the rocky face. The three larger Gulls, the two Blackbacks and the Herring-Gull, were also present, attending to their broods or eggs on the grassy ledges at a moderate elevation, and Black Guillemots were observed entering the dark cracks and crevices, as well as making their way beneath the large slabs of rock lying on the sloping patches of rank vegetation, scattered here and there along the whole range of cliffs. The positions on the rocks in which the Guillemots, Razorbills, and Kittiwakes had taken up their breeding-quarters were the same as those observed by Mr. MacGillivray during his wanderings in the Northern Highlands. In vol. ii. of his 'British Water Birds' (p. 321) he states:—"When the cliffs are high, and other birds breed upon them, the Guillemot occupies a zone above the Kittiwakes and below the Razorbills; but when the latter are not present they disperse over the face of the rocks."

The accuracy displayed by this excellent observer in all his descriptions contrasts strongly with the errors often so obvious in some of the best-known and most expensive works on British Birds. In Gould's 'Birds of Great Britain' this species is depicted at its breeding-quarters, and although the plate is beautifully executed and the birds correctly drawn and coloured, the site for the reception of their eggs, a bright green grassy ledge, is one on which I never observed them making an attempt to rear their young. The rocks in every instance where I have examined their haunts were white with the droppings of the birds, and entirely free from any but the most scanty vegetation. The six eggs represented in the plate all exhibit a pale blue ground-colouring, and correspond almost precisely in the black markings: judging from personal observation, I am of opinion that the same tint is seldom, if ever, seen on all eggs in view; indeed I doubt if a couple could be found exactly alike on any range of cliffs. No one can deny that Gould's magnificent work gives far the best representation of the plumage of the feathered tribe; the plants and general surroundings also are wonderfully true to nature; still all who have studied the birds while in life and at their native haunts must regret that several such inaccuracies as those referred to are to be found in its pages.

No eggs vary to such an extent as those of the Common Guillemot; the ground-tint is occasionally dark or light blue or green of several different shades, speckled and streaked with black. At times, the whole shell is almost a pure white or pale yellow with a few clouded grey markings; the majority, however, are suffused with brown and grey of various hues, the markings consisting of blotches, spots, and scrawls of darker shades of the same colours. To describe these curious eggs accurately would need an endless series of coloured plates.

Though the eggs of the Common Guillemot differ to such an extent, I have good cause for believing that each bird always lays those of the same colour and markings. Towards the end of May 1865 I removed three eggs from a small ledge on the north side of the Bass Rock, and on visiting the spot about ten days later found three more in the same situations, and so exactly like the former as to be undistinguishable. Again returning a fortnight later, three more, similar in colouring of the shell and corresponding almost mark for mark, were obtained. No other Guillemots were breeding within twenty or thirty yards of that portion of the rock, and though I frequently examined the spot from the summit through the glasses, no more than three pairs frequenting the ledge were observed.

When the young are fit to leave the ledges on which they were hatched, the evidence of many trustworthy individuals goes to prove that they are either carried down on their parents' backs or lifted in their beaks by the neck, and so transported to their natural element. For my own part, I can offer no opinion on the subject, having, in spite of many attempts in different localities, failed to witness the operation. Numbers of juveniles have fallen or been knocked over from a height of one hundred up to two hundred feet while I have been near at hand, but unless striking against a projecting rock in their descent, they never, so far as I was able to ascertain, sustained the slightest injury. In the summer of 1865 I reared one that dropped considerably over two hundred feet from one of the highest ledges on the west side of the Bass Rock, and only missed the side of our boat by a couple of feet. This interesting little stranger lived for three or four months, apparently contented and happy, roosting every night in the coal-cellar, where he climbed to the highest pinnacle of coal, evidently imagining himself on his native Bass Rock. In the morning he generally appeared as a Black Guillemot till restored to his natural colour by a salt-water bath in a large washing-tub. Like most favourites, he came to an untimely end; one morning after a heavy thunder-storm he was found dead at his accustomed resting-place on the top of the coals, having perished from the combined effects of wet and cold, caused by the bursting of a water-pipe over his head—dying where he sat, rather than make an attempt to quit his post.

In 1874 we took several young ones from the Bass Rock between the 10th and 13th of August; these appeared to be the latest stragglers left on the rock. The whole party proved exceedingly easy to rear, taking the food offered to them with the greatest avidity; they evinced a decided partiality for herrings, though when these were not procurable they were forced to put up with whiting. While at Canty Bay the little divers were confined in one of the boathouses, and we treated them daily to a swim in the waters of the Firth. Having been carried down to the shore they were flung some fifteen or twenty yards—as far as the men could pitch them—out into the sea, when they immediately started back, swimming, fluttering, and flapping, to the best of their ability, uttering all the time their monotonous and plaintive cry "*quilly, quilly, quilly, quilly,*" and appearing in the greatest trouble till, having been rolled ashore by the breakers, they succeeded in scrambling up the shingle, and were safe back in charge of their captors. While sunning themselves on the rocks to dry their ruffled down and sprouting quills, their greatest delight was to crawl up the sleeves of the boatmen's rough pea-jackets and endeavour to seek repose; when shaken out again they resented the injury by giving vent to a succession of shrill notes of lamentation. These birds were taken south to Brighton, and lived in confinement for several years; there proved to be no necessity for pinioning them or cutting their wings, this species being unable to rise from land unless assisted by the drop from the cliffs to which they resort for breeding-purposes. At sea they usually flap over the water, striking the surface with their feet for from ten to twenty yards, or perhaps double that distance*, before rising on wing, and the pond to which they had access being only some five-

* The distance they flap over the water before rising on wing depends on the amount of food recently consumed; if light they fly off with comparative ease, though when crammed with fish they are often incapable of rising from the surface, and after fluttering a short distance usually attempt to seek safety by diving.

and-twenty feet across, entirely precluded them from gaining sufficient impetus to top the fence or the willows surrounding the enclosure. Eventually accidents, rather than natural death, cleared off the greater number of these interesting little strangers; their companions the Gannets were responsible on more than one occasion, the last survivor being swallowed by one of these voracious birds, and only thrown up again on the following day after decomposition had partly set in*. Such mishaps I learned have also befallen this species in the Zoological Gardens: the keeper who looked after a pair informed me, at the latter end of January 1874, that these birds were now and then moved to the enclosure in which a Pelican was confined, in order to enjoy a larger bath than their own domicile afforded; all had gone well at first, but at length the rightful owner resented the intrusion to his pond and snapped one of the offenders, and it was only with the greatest difficulty that the unfortunate was rescued from the pouch beneath his beak. While in confinement these birds appear exceedingly attached to one another, roosting and resting side by side, sitting on the ground together with their heads buried in one another's plumage, and occasionally preening the feathers on the throats of their mates. While this little amusement is going on, the birds utter a succession of low murmuring notes, the one operated upon evidently being gratified by the attention and returning the compliment on the first opportunity.

Eight or nine years ago I turned out half a dozen adult Guillemots in one of the large tanks at the Brighton Aquarium, and these birds soon proved one of the greatest attractions to the place. The last time I visited them they were busily engaged in diving alongside a large skate that was slowly swimming up and down the tank. Some air-bubbles escaping from the gills of the fish appeared to afford great amusement to the birds, as first one and then another dived down and caught them before they rose to the surface. The old birds I presented have all died in due course, but their places have been filled by others procured by the company from various parts of the British Islands.

By the introduction of Guillemots and Razorbills to the tanks of the Aquarium, an excellent opportunity was obtained for ascertaining the amount of air conveyed around their feathers when below the surface of the water. While diving in pursuit of food, these birds make use of their wings as if on flight, and also propel themselves with their feet, dashing forward at times with great speed, and occasionally moving slowly with a jaunty undulating motion. If closely watched while passing round the tank, the whole of the plumage of the body as well as that of the head and wings will be seen thickly coated with air, which glistens in the light with a glassy appearance. As they proceed, a continued stream of silvery bubbles bursts out from their feathers, and lights up the track they follow with a brilliancy that would scarcely be imagined. The plumage of the back may be seen opening as the air escapes, and the bubbles stream up the feathers till they reach the extremity of the tail, when they pass upwards to the surface of the water. The greater number of these luminous bubbles are somewhat less than a pea, and others, larger and more attractive, vary in size from threepenny or sixpenny pieces up to a shilling, and assume all manner of shapes and forms when released from the feathers and rising towards the surface.

Though considered a species and figured as such by several authors in former days, the Ringed, or Bridled, Guillemot (*Uria lacrymans*) has now been pronounced by the greatest authorities to be merely a variety of the common form. With this decision I certainly agree, having carefully watched the two forms paired at various breeding-stations. At the Fern Islands the Ringed Guillemot is exceedingly numerous, some twenty or thirty often being in view at once on the Pinnacles; the confusion here, however, is so great when an alarm is given that it is seldom possible to make any accurate observations as to the pairing of the birds. On the north face of the Bass Rock one may remain concealed on some of the larger ledges, and by the help of powerful glasses inspect all the pairs of Guillemots within view while engaged in conveying food to their young. In August 1874, while watching for this form, I marked an adult with the ring and bridle well developed brooding over a downy youngster, and having procured the ropes secured the juvenile, who exhibited not the slightest difference

* These facts are fully referred to in the paragraph commencing line 7 on page 11 of the Gannet.

from others taken on the same and two or three adjoining ledges. On the immature plumage being assumed later in the autumn not the slightest signs of the white markings round the eye were exhibited, and though the bird survived in captivity for three or four years, no change took place. The following year an old bird was caught with her young one on the Bass by the men who look after the roping business to secure the Gannets, and forwarded safely to Brighton. Both thrived well in confinement, and the adult showed the ring till her death about four years afterwards; the young one, however, was not to be distinguished from several others procured at the same time. In the Channel I have seen these birds very plentiful on several occasions; numbers were passed sufficiently near to be identified, and a fine pair in perfect breeding-plumage obtained, eight or nine miles off Brighton, on the 23rd of April, 1884, the water at the time being as smooth as glass and affording excellent opportunities for observation. In the perfect winter dress the white ring and bridle are still conspicuous, a narrow line of dull black enclosing the bridle. A bird exhibiting this stage of plumage was shot off Rottingdean on December 17th, 1878.

BLACK GUILLEMOT.

URIA GRYLLE.

The true home of the Black Guillemot in the British Isles is evidently along the northern and western coasts of the Highlands; I failed to detect any breeding-stations further towards the south-east than Duncansby Head, though there are doubtless other portions of the dull red cliffs of Caithness, overlooking the North Sea, to which they resort. I could gain no information from the two oldest fishermen at Canty Bay, John Kelly and Andrew McLean, both acquainted with the Bass from their boyhood, that this species had bred upon the rock, though the fact has been recorded by more than one writer. Unless the interior of the caves or the ruined buildings among the fortifications were made use of, the formation of this rocky island does not seem adapted to their requirements, as I never observed their eggs on open and exposed ledges, and no other accommodation could they possibly have secured in the face of these cliffs. At the May there is, so far as I have been able to ascertain, good evidence that some years back a few pair frequented certain portions of the island and regularly reared their young. While in pursuit of Ducks along the shores of Gullane Bay, in the Firth of Forth, after a strong north-east gale and a heavy sea, in January 1861, I discovered two dead birds of this species in winter plumage lying at the high-water mark among the weeds cast up by the tide. In 1868 and the following year I often drove in early spring through Sutherland and Caithness to the north, and repeatedly remarked small parties at sea actively ducking and diving in the waves off the coast about Golspie, Helmsdale, and Dunbeath. A Black Guillemot, in almost adult though worn and faded plumage, was obtained just outside the rock on which the outer lighthouse stands at the Fern Islands in May 1867; no others were seen, and I was unable to ascertain from Darling, the egg-collector, that they were more than uncertain visitors at the present time to that part of the coast, though supposed to have been summer residents in former days.

I passed some time in May 1868 in exploring the west coast of Ross-shire, and in the deep cracks and crevices in the face of the red cliffs, or the detached rocks near the Stack Buie, detected with the help of the glasses numbers of eggs of this species. Concealed among the huge blocks of fantastically shaped stone encumbering this wild and rugged shore, I was enabled to watch the birds fly in from the Minch and make their way towards their gloomy quarters: after alighting on the ledges they usually paused for a few moments, gazing eagerly around, then stooping forwards, rapidly disappeared in the shade below the overhanging rocks. My attention was particularly attracted by the rapid movements of these birds and the speed with which they made their way among the stones or over the rocks, their actions contrasting greatly with the shuffling gait of the Common Guillemot, which invariably drops from the air into the ledge it frequents, in close proximity to its egg or young. This species also possesses the power, not bestowed on its relative, of rising from a flat surface; the latter, as all are aware who have studied their habits in a state of nature, gain no assistance from their wings till a considerable drop below the ledge from which they spring has been effected. On the island of Fura, and on other spots along the coast, I remarked that the breeding-quarters of these birds were often situated under the large slabs of stone scattered over the ground, often at a distance of twenty or thirty yards above high-water mark. Though taking little notice unless their haunts were too closely inspected, when an alarm was once raised they usually beat a speedy retreat, and on making their appearance from beneath the shelter at once took wing, rising with almost the ease and rapidity of a Partridge. The elevation in the fissures selected by this species for breeding-purposes is generally only sufficient to allow the birds to make their way towards the eggs, which are usually located at the distance of three or four feet from the entrance to the cavity. The crooked sticks commonly carried by Highland shepherds proved most useful in extracting the few I needed as specimens; unless fresh laid, there were invariably two eggs.

In a deep chasm among the rocks at Duncansby Head I noticed a pair of these birds feeding their young beneath a large grey moss-grown slab of stone lying on a slope green with grass and ferns, on which a pair or two of Gulls were also nesting. The little Divers were busily engaged in procuring food, and arrived on the scene repeatedly during the hours I remained on the summit of the cliffs watching the swarms of sea-birds passing within view of my position; at various times one or the other stopped to rest or clean its plumage on the slope at the edge of the precipice before dashing down to the waves. I remarked that the visits paid by these Guillemots to their domicile occasionally attracted the attention of a fine old Great Black-backed Gull, sitting demurely on her eggs within a few feet of their quarters; now and then she turned her head in a threatening manner, but took no further notice.

PUFFIN.

FRATERCULA ARCTICA.

While conversing with the natives in a remote district of the Outer Hebrides a few years back, I listened, among other yarns, to a strange account that was given concerning the manners and customs of the Puffin, in the veracity of which it was evident that my informants placed the most implicit confidence. The men stated that this quaint-looking sea-bird (to which they gave the name of Tammie Rookie) was the most knowing of all feathered creatures, declaring that after a winter's absence on the open sea large flocks appeared at a certain date (never a day out) on the saltwater lochs, and after remaining for a time in the vicinity of the land, the whole assemblage betook themselves with the same regularity to their accustomed breeding-haunts on the adjacent rocky islets. It was the custom among the fisherfolk and crofters to visit the quarters to which these birds resorted every spring in order to obtain a supply of eggs; and so well aware was the astute Tammie Rookie of the annual depredations from which he suffered, that on the approach of the robbers he immediately entered the small burrowed domicile in which his only treasure was concealed and at once rolled out his egg—this action being looked upon as a mute and irresistible appeal that his life might be spared *.

From the remarks in several works it will be seen that the inhabitants on other parts of the northern coasts have much the same belief regarding the regularity with which this species seeks the land on the approach of spring. An early date is given by certain writers for the arrival of the Puffins in the Minch and other waters along the west coast; I noticed, however, on one occasion that these birds did not show on several of the saltwater lochs off the coasts of Harris and Lewis till the 3rd of May, when small parties, numbering from six or eight up to a dozen, made their appearance.

The breeding-stations of the Puffin are so well known that it is needless in these pages to give a list of those I have visited. Assemblages of countless thousands are described as occurring along the wild and unfrequented coast-line of the Western Highlands. I remarked a few places where they had taken up their quarters, but have never had sufficient time to explore only a very small portion of the range of rocks they frequented. Though the multitudes that collect in these localities are doubtless far greater—so many that I should be unwilling to hazard a guess at their numbers—they might have been seen a few years back congregated in thousands at certain hours about the island of Craig Leith in the Firth of Forth. On several occasions during the latter end of July 1865, while brought up fishing about a quarter of a mile to the east of this rocky islet, its whole outline stood out clearly defined as the sun gradually disappeared, and a most animated scene was witnessed as long as daylight lasted. Clouds of Puffins were flying round; a few alighted from time to time on any vacant spot, every ledge on the rock appearing a living mass of birds, while the sea for several hundred yards around was

* A bird seized by the hand in its burrow must, while struggling to escape, have kicked out its egg, and in this manner given rise to this strange idea.

thickly covered by large flocks floating quietly on the surface or rising occasionally on wing. In the twilight hours Puffins invariably exhibit the greatest activity; it is only about sunset or shortly after daybreak that I have noticed these immense gatherings in this part of the Firth.

A wounded Puffin always attacks his captor with the greatest determination, though the bite he is able to inflict is by no means so severe as might be expected from the form of his singular bill. While procuring specimens in summer plumage off the island of Craig Leith, in June 1865, I succeeded in stopping half a dozen (all that were needed) at a single shot. The majority of the victims proving wounded, an attempt was first made to secure a lively cripple struggling on the surface; two or three others that had dived, rising as this bird was seized, fixed on my arm, and were lifted into the boat without further trouble, all signs of the nips they administered rapidly disappearing. It is evident that several writers on British ornithology, who have described the beak of this species as hard and sharp and capable of inflicting a severe wound, can never have procured the birds themselves and examined them at once, as this is not the fact. In life the edges of both mandibles are broad, soft, and fleshy, and it is only some time after death, when all the soft parts have dried up and contracted, that such an idea could be formed, as the bite the poor wounded bird inflicts, if he happens to grip your hand when you lift him into the boat, is scarcely perceptible.

This quaint-looking species resorts to a variety of situations for breeding-purposes; though well able to excavate a hole for themselves in soft mould, they occasionally fraternize with the rabbits, or at times (probably owing to the effect of the guano) evict them from their burrows. To ascertain the numbers that rear their young among the ruins of the old fortifications on the Bass, the rock must be visited shortly after daybreak or at dusk. In the lower dungeons I have watched these birds emerging from cracks and crevices among the masonry, as well as from beneath the fallen blocks encumbering the passages and adjacent ledges. Hundreds in early morning, when no cause for alarm has arisen, may be seen resting quietly on the projecting stones of the building or the slabs below the openings by which they gain an entrance; during the day not a Puffin is to be observed, so I suppose they may be considered somewhat nocturnal in their habits.

To the best of my knowledge, the intermediate stages of the Puffin have been overlooked by the majority of writers; neither has any opinion been given as to the age at which the bird assumes all signs of maturity. Few of this species appear to approach the land while exhibiting the intermediate stages unless driven ashore by the buffetings of protracted storms. The first immature Puffin that came under my notice was obtained in a dying condition off Whitby on the Yorkshire coast in July 1862: the second was secured alive off Brighton on the 1st of July, 1874; this specimen was turned out in the garden, and taking up its quarters in a small pond, thrived well for over a year, no alteration having taken place in the colouring of the soft parts or plumage at the time of its death. While fishing off the Bass, an immature bird was taken in the landing-net floating helplessly past the boat after a gale in October 1874; another in precisely similar dress, and corresponding in the colouring of the soft parts, was obtained off Brighton during the first week of April 1879, the last in this stage that came under my observation being captured by a fisherman on the 13th of February, 1883, a few miles off Lancing. The notes taken down descriptive of the tints of the soft parts of these birds show a great similarity, little if any distinction being apparent between those obtained during summer, autumn, or winter. In order to give a correct representation of the beak of this species, it is necessary that a sketch should be taken immediately after the specimen is secured. The bright tints on the soft portions of the mandibles, that render the adults so conspicuous in summer, rapidly shrivel up and fade; a change also takes place with birds procured in winter, the lower mandible contracting to a considerable extent and the markings gradually disappearing.

An adult Puffin in the full breeding-plumage is represented in the Plate; the colouring of the soft parts, which may possibly be found to differ from the figures in certain works, was taken from a sketch made within a few minutes of the death of the bird, which was shot at the Bass on the 10th of August, 1874. With regard to the age of the juvenile, I can offer no opinion, never having succeeded in keeping this species in confinement for any length of time: the specimen whose portrait is given was caught off Brighton on July 1st, 1874; I remarked that in every particular it corresponded with others obtained at various dates during autumn and winter. It is needless to state the colours on the beak, feet, or legs of the adult; and those of the immature bird are also plainly shown in the Plate. On the 13th of February, 1883, I examined a specimen, the soft parts of which may be described as follows:—Iris dark brown; no circle round the eye. Upper mandible a red-brown at the base, turning into a darker brown at the point; lower mandible a livid slate tint near the base, turning into a warm brown towards the point. Inside of mouth and tongue pale yellow. The legs a livid blue at the knee-joint; yellow tinge down fore part of tarsi and a livid flesh on the back part. Webs a dusky olive-yellow; toes yellow, with the exception of the outer on each foot, which was of a pale livid flesh-tint; nails black.

Though various other names are bestowed on this species, it is only as Tommie Nories on the east coast of Scotland, and Tammie Rookies on the west, that I have heard these birds spoken of by the natives of the localities in which they abound.

RAZORBILL.

ALCA TORDA.

A CONSIDERABLE diminution in the number of Razorbills frequenting some of the breeding-stations on our north-eastern coasts has taken place of late years; at the time of my last visit to the Fern Islands, only one pair of these birds had taken up their quarters for the season, though I learned from the egg-collector that they had been formerly abundant during summer on various parts of the rocks. On the Bass also I remarked a great falling off in numbers; the upper ledges on the north-west face had been formerly a favourite resort for this species, but on searching for young birds in the autumn of 1874, in order to rear them in confinement, their old haunts were found to be entirely deserted. Along the rocky cliffs about Duncansby Head I found Razorbills very numerous at the time of my visit in 1869, the higher ledges in many parts appearing to be entirely tenanted by this species. At a point near the east end of the rocks I had crawled to the grassy edge of the precipice to look down upon the sea-fowl, and on stretching over came face to face with three or four pairs whose quarters were within a couple of feet of the summit of the cliff. I was almost as much startled as the birds: my intrusion on their haunts, however, caused but a momentary alarm; one or two shuffled up against their neighbours with angry croaks, but soon settled down again in their accustomed manner, paying little or no attention to my presence.

While shooting at sea off the coasts of Kent and Sussex I frequently remarked during winter that previous to the setting in of stormy weather in the Channel, Razorbills were exceedingly restless, immense flocks of these and other Divers being seen on wing making their way either east or west for several hours. No general movement to any distance appeared to be taken; within a day or two the stream of birds would probably be seen following an opposite course.

At times, when the fry of fish are plentiful and making their way along shore, I have seen these birds perfectly crammed with food, snapping up the glittering morsels by merely dipping their heads below the surface without the exertion of diving. During fine weather with a light northerly breeze in early spring I have repeatedly watched a pair or two greedily stuffing themselves from our boat-house at Shoreham, and ascertained that their prey was secured in the shallow water close to the shore, and occasionally even among the tiny breakers rippling over the shingle.

I have observed Razorbills exceedingly numerous even in the summer months inside the sands off the Norfolk coast. So early in the season as August 9th in 1873, an entry in my notes referring to the subject occurs; though swarming within a mile or so off the shore, there was not a bird to be seen on the open sea outside the sands. It is not often that Divers of this family are taken on board the Trinity vessels; a Razorbill, however, struck the lamps of the 'Newarp' floating light-ship in March 1873.

Razorbills are termed Marrots by the writers of several ornithological works, the name being probably derived from a supposed similarity to the sound of their note. On many parts of the coast they are confused with Guillemots by the fishing fraternity, and are known as Willocks or Willies. The men composing the crews who assisted me at my work often begged for a few to be shot when the birds were abundant in the Channel, stating that they proved excellent when baked, after having been skinned and laid in fresh water, by which means the fishy flavour is removed.

I am not acquainted with any breeding-stations of this species in the south of England, within many miles of the Sussex coast, that are resorted to at the present time. That such still exist, however, is evident, as a fisherman who was working his shrimp-net over the sands near Shoreham on the 9th of August, 1883, captured in the shallow water a young one that had strayed some distance from the old bird. So far as I could judge (my examination having unfortunately been a "post mortem"), the downy juvenile was about ten days or a fortnight old. The man who secured this variety, knowing that I would like to attempt to rear it, carried the little stranger home and sent word to me at Brighton. On arriving and making inquiries, I learned that his wife, having been wearied by its continued cries for food, had put an end to its existence, declaring that she could not bear to hear it "calling for its titty."

CORMORANT.

PHALACROCORAX CARBO.

FROM Caithness to Cornwall Cormorants may be met with at one season or another, their haunts being either in the cliffs and rugged rocks overhanging the sea, or about the mudflats surrounding the estuaries of the low-lying portions of our shores. These birds also make their way inland, frequenting rivers and occasionally resorting to the broads and meres, evincing a partiality for those waters well stocked with fish. I often watched a bird or two perched on the posts marking the course of the river across Breydon mudflats near Yarmouth, and others were noticed on the beacons that guide the steamboats on Loch Ness. An old ferry-man living at Dochfour, near the east end of the loch, who often rowed my boat while fishing, was loud in abuse of these birds, which he termed Crans, declaring that they consumed immense quantities of trout; and doubtless the opinion he had formed concerning their destructive habits was correct. At the time of my last visit to the Fern Islands, in 1867, there were about one hundred pairs breeding on a low rocky island, the highest nest being only about twenty feet above high-water mark, and the whole easily accessible. I learned from Darling, the egg-collector, that these birds usually change their nesting-quarters every season. The eggs of this species are sold at the neighbouring villages or to chance purchasers at four a penny *. On the Yorkshire coast, a few miles north of Whitby, I noticed a considerable number, in the summer of 1862, breeding high up in the face of the cliffs, the ledges on which the nests were placed being perfectly free from intrusion unless ropes were brought into use. The numbers of Cormorants visible on the rocks along the wild and dangerous Cornish coast are sure to attract attention; in the vicinity of all the bays and coves they appeared especially abundant, fishing continually and flying up, to rest or dress and dry their plumage, to the face of the cliffs or the detached rocks that offered security from the breaking surf. As I believe the young have not been figured by any author, a description of the soft parts of an immature bird shot near Lamorna Cove on November 5, 1880, in its first autumn plumage, may not be out of place:—Iris dirty green, circle round a dusky yellow; upper mandible a dull dark brown, lower a pale yellow tinged with green; gape yellow. Legs outside dark, almost black, inner side pale yellow; webs and toes dusky, joints of toes darker; nails black.

The season at which the male Cormorant attains the milk-white neck is doubtless early in the spring, and the bright feathers are in all probability speedily shed. During all my travels in the north, I have seen but few birds displaying this very conspicuous colouring. On the 14th of April, 1869, I noticed a Cormorant with a perfectly white neck resting on the sand-banks to the east of the channel that forms the outlet from the muddy harbour known as the Little Ferry, near Golspie. Though there was little hope of securing this handsome specimen, I determined to make the attempt. A heavy sea was rolling in the channel, with a strong wind blowing from the south-east, as we swept out from between the banks, and so much water broke on board that it was necessary to run aground to save sinking the

* Full particulars concerning the egg-business at the Fern Islands are given under the heading of the Arctic Tern.

punt and losing our gear. The bird took little notice of our mishap, and might easily have been secured had the wind and weather only been favourable. On the Tain sands, on the opposite side of the Dornoch Firth, I also noticed one in the same state of plumage, but failed to obtain a shot, owing to the state of the tide. At none of the breeding-stations visited in May and June was I able to detect a single bird exhibiting the white feathers on the neck to any extent, and in all probability this conspicuous colouring is discarded before the nesting-time. Repeated observations also led me to believe that it is only the oldest birds that assume this state of plumage. The most perfectly marked white-necked specimen that has come under my notice was in the Inn at Garve, on the road between Dingwall and Gairloch, and I learned that the bird had been shot on the freshwater loch near at hand.

Though it has been asserted that the flesh of Cormorants is useless as an article of food, these birds, while in the immature state, are by no means unpalatable when served up by those well used to their work. One spring, while stopping at Canty Bay, I happened to enter the kitchen of the inn when Adams the landlord, who also rented the Bass, was at dinner, and at his request I sat down and was helped to (what he was pleased to term) a plate of hare-soup. It was undeniable that the soup was excellent, but it is doubtful if it would have been approved of had I been aware of the fact that it was made from certain portions of two Cormorants and a Shag shot near the Bass a few days previously. I remarked, while living at North Berwick and Canty Bay, that though using the Bass as a roosting-station in great numbers during autumn, winter, and early spring, no Cormorants ever bred upon the rock. Some of the old fishermen, who had looked after the birds for over fifty years, also informed me that, to the best of their knowledge, not a nest had ever been built by this species on the Bass.

Cormorants are well known to select singular situations on which to alight; a bird has now and then been seen on the spire of the highest church at Yarmouth, and also more than once on a brewery-chimney in another part of the town. On the 16th of September, 1868, I watched an immature bird, after several failures, obtain a footing on the tail of the gilt cock on the summit of the town-hall at Tain, in Ross-shire; here it remained, balancing itself in an uneasy manner with no little difficulty, till fired at from the High Street, when it at once made a straight course for the Firth. While gunning on the north-east coast of Scotland in March 1869, a Cormorant made an attempt to settle on our punt as we dropped quietly with the ebb-tide down a channel in a muddy harbour, just as daylight was closing in.

The nests of Cormorants are placed, as previously stated, on ledges in the face of lofty cliffs and also on low rocky islets at but a slight elevation above high-water mark; in two or three localities these birds also resort to the islands on inland lochs, constructing their nests on the open ground. Some twenty years ago they often built in trees on both the north and south shores of the Dornoch Firth (Scotch firs, I believe, in every instance); but on making inquiries a few months back as to the numbers that now frequented the spot, I learned that no birds had been seen in the district for some time. The nests I closely examined at the Fern Islands were large and roughly constructed, composed for the most part of weather-beaten branches of trees and dead sticks, the greater portion of the materials having evidently been picked up at sea, the bark being worn off from long immersion in the waves or washing along the shore. The stalks of the largest tangle were also intertwined, apparently with the view of holding the structure together, the lining being invariably composed of various kinds of the smaller seaweeds, with now and then a few tufts of rough grass.

SHAG.

PHALACROCORAX GRACULUS.

From personal observation I can say but little as to the distribution of the Shag over the British Islands; it has come under my notice only on the north-east coast of England, in the Firth of Forth, and around the northern and western coasts of the Highlands. Unlike Cormorants, these birds do not appear to make their way inland; I never observed even a single individual on any of the Scotch lochs, the Norfolk broads, or the Sussex rivers, all of which are favourite resorts of the nearly allied species.

While collecting specimens near the North Point, in the west of Ross-shire, at the end of May 1868, I was informed by the keepers that numbers of Cormorants bred in the caves and crevices of the rocks surrounding the Black Bay, and also near the Ru Rae; on reaching the spot, however, I soon ascertained that the whole were Shags. The scenery on this part of the coast is wild and desolate in the extreme—the deep chasms and rifts in the cliffs, as well as the larger caves frequented by the Shags, being dark and gloomy, forming a most fitting residence for these uncanny-looking birds. The nests and their occupants were in many instances scarcely visible in the dim light that penetrates to these recesses, the flash of an eye or the movement of a head usually attracting attention to the presence of the inhabitants. Several nests were placed at no great height in the caves, and, standing on the shoulders of a couple of sturdy Highlanders, I was enabled to inspect their contents, which consisted of both eggs and young, the latter for the most part newly hatched. The breeding-stations on the Pinnacles at the Fern Islands appeared to be almost deserted at the time of my last visit, in the summer of 1867, only a single pair being seen about the islands. Though formerly a favourite resort of these birds, the east and west caves at the Bass Rock in the Firth of Forth are but seldom tenanted by them at the present day.

The nests I examined on the west coast of Ross-shire were built with large heather-stalks, together with a quantity of smaller twigs, a few stems of ferns with those of other plants were also intermixed, a warm and closely interwoven cup-shaped lining of coarse strands of grass and rush forming the interior*. The cradles constructed by this species are, as a rule, of far less bulk than those of the Cormorant; the juveniles on breaking the egg-shell are naked, black, shapeless monstrosities, covered in due course by a hairy down of dark grey, almost approaching black. The immature in the first stage are arrayed in a light-breasted plumage after the manner of Cormorants; having failed, however, to make satisfactory observations on either species while kept in confinement, I am unable to form an opinion as to the age at which the full adult plumage is assumed, though doubtless it is not put on till the third or fourth year. The beauty of the plumage of an adult can scarcely be judged by stuffed specimens or plates; the bird must be viewed immediately after death, or the wondrous hues of bronze and green on the feathers of the back, which seldom retain their natural gloss for any length of time, can never be seen to perfection.

* I refer to the composition of these nests somewhat minutely as the young are depicted in Gould's 'Birds of Great Britain,' the most beautifully illustrated work that I have ever seen, on a few strands of green seaweed and moss lying on an exposed rock a few feet above the level of the sea. The plate, however, gives by far the best representation of the old bird and young to be seen in any ornithological publication, and but for the singular error referred to would have been as near perfection as can possibly be attained.

GANNET.

SULA BASSANA.

At one season or another Gannets are to be met with on all the seas that surround the British Islands. The breeding-stations to which they resort are but few in number, and during the summer months the old birds seldom stray to a greater distance than from fifty to one hundred miles from their quarters. According to my own experience, the rocks they frequent for nesting-purposes are deserted soon after the young are able to take their departure, the latest stragglers, with but few exceptions, leaving their summer haunts about the first or second week in October. For the following four or five months the birds make their home on the stormy ocean, seldom returning to the land unless blown ashore when weakened by the buffetings of continued gales. As winter draws on, flocks of Gannets of various ages, the old birds predominating, show themselves in attendance on the shoals of herrings in the North Sea, gradually following the fish towards the south. My visits to the neighbourhood of the haunts of these birds on the west coast of Scotland having been made, on almost every occasion, during spring or summer, I have had but few opportunities of studying their movements along that coast in the autumn.

The Bass Rock, situated a couple of miles at sea in the Firth of Forth off the coast of East Lothian, is the only breeding-station of this species with which I am well acquainted. There are in all, on various parts of Great Britain, some half dozen other spots to which the birds resort during the summer; wind and weather, however, have invariably combined to frustrate the attempts I have made to visit any of these quarters.

The extent of ground occupied by the colony on the Bass has considerably diminished of late years, many of the stations formerly resorted to by from forty or fifty up to one hundred pairs on the more accessible slopes near the summit being now entirely deserted. The constant interference from sightseers, as well as the repeated robbery of their eggs by irrepressible tourists, gradually drove the birds from such exposed nesting-quarters. The falling-off in the numbers in consequence of the desertion of these stations is not so great as is usually supposed. I have remarked a considerable addition to the nests on the small ledges on the north-east face and also near the east cave: these spots are almost inaccessible, even to the regular egg-takers (the ropes with which they are supplied being none of the best); and no young being collected, the existence of these birds is not reckoned by the tenant of the Rock.

It is seldom any Gannets are seen in the neighbourhood of the Bass during the winter, and it is not till early in March that they begin to collect. Rough and stormy weather, however, occasionally causes them to take their departure again for a time, but by the end of the month there is generally a considerable gathering. On one occasion (1867) an egg must have been laid as early as the last week in March, since a young bird was hatched on the 10th of May; this was fully a month sooner than the usual time of laying, many of the birds only commencing their nesting-operations after this early youngster had made his appearance.

There are reckoned to be about three acres of pasture * on the upper portion of the Rock; and up till some twenty years ago the birds made use of several open grassy slopes above the buildings on the south side and also on the summit of the cliffs towards the north. The nests on these spots have now disappeared; but at the time of my last visit a very moderate climber might, without the slightest risk, explore many of the ledges still resorted to. In the early part of the season, when the birds commence laying, a small number of eggs are annually taken. A few are bought by collectors at the price of sixpence each; and I have heard the boatmen informing credulous sightseers that a score or so were sent regularly to London, as Royalty was pleased to approve of them for breakfast. I cannot vouch for the truth of this statement, having omitted to make any inquiry. But a single egg is deposited; and the men in charge of the Rock declare that, unless this is removed the morning after it is laid, the female will make no further attempt at nesting for that year. In a few instances, perhaps once or twice during a season, I have remarked a couple of eggs. It is usually supposed when this occurs that two pairs have made use of the same nest or the eggs have been moved by visitors. The bird engaged in the duty of incubation invariably, before sitting, spreads the webs of one foot carefully over the egg, next places the other foot as closely as possible over the first, and then drops onto the nest. I have watched this habit in the wild state and also with those I kept in confinement.

The Bass is a busy sight about the latter end of June, by which date young birds are to be observed on most parts of the Rock. While small and uncovered with down the nestlings are seldom left unprotected, one or other of the old birds remaining constantly on guard. It is amusing to watch the greetings that take place between the pair when one returns after a lengthened voyage in search of prey. The whole of the food carried to the Rock is swallowed when caught, and afterwards disgorged either on or close to the nest. The pieces of fish that are thrown up, whether whole or in fragments, are again greedily devoured with much clamour by both parents. When their meal is partially digested the young bird is fed. The nutriment supplied to the nestling in its earliest stages is in either a soft or almost liquid state: the old bird, stooping forward, stretches down its head sideways, and opening its jaws to the fullest extent, completely conceals its diminutive offspring; the young one then helps itself to the preparation ejected into the gullet of the parent.

Though the food is usually picked up either in the Firth or a few miles at sea, I have noticed adult Gannets on two or three occasions fishing off the Dogger Bank in the latter part of June. It is most probable that these birds were from the Bass, though the distance could be little short of two hundred miles. While fishing or cruising in the North Sea, between twenty and thirty miles off the Rock, I now and then observed small parties returning from their fishing-excursions. When a supply can be obtained in the Firth, the birds are to be seen dashing down incessantly, and repeatedly making their way to the Rock to discharge their cargo. In the half-rotten mass of fish on some of the ledges it is easy to recognize the result of each catch laying in the shape in which it was disgorged. Half a dozen herrings or three or four mackerel may commonly be noticed neatly and compactly wedged together. In some instances the heads are partially digested; this probably occurs after a protracted journey in quest of food has been undertaken. As many as half a dozen of such parcels of fish are at times to be counted round a single nest; though considerably larger, they strikingly resemble the bundles of dried sprats exhibited on the counter of the fishmonger. There are also among the cracks and crevices of the rocks countless portions in a state of decomposition. The whole of the refuse forms a most uninviting and repulsive spectacle, while the disgusting odour that assails the nasal organs can be more easily imagined than described. As the young increase in size, and the nests and ledges become thoroughly saturated with guano, the effect is considerably heightened; the stench that arises from the steaming nests under the scorching rays of the summer sun after a wet morning is positively overpowering.

* Some years back a score or so of sheep used to be sent out to feed on the island. A pulley fixed among the old fortifications on the south face of the rock, where the cliffs are perpendicular, was employed to land the animals, lifting them above the narrow and slippery footway up which visitors are forced to make their way towards the summit.

I have at times been forced to beat a speedy retreat when attempting a visit to some of the breeding-stations on the north face of the cliffs.

The young Gannets at the age of a month or six weeks (their black faces surrounded by tufts of down which strongly resemble white nightcaps) are exceedingly comical, though for the most part ill-tempered and peevish little tyrants. When first they waddle a yard or so from their nest, it is most commonly to pitch into some smaller and more helpless infant, which is not unfrequently seized by the back of the neck and shaken in the most pitiless manner. Though the injury they are capable of inflicting on one another is slight, their battles are often attended with fatal results; one or other, or occasionally both, of the combatants lose their balance, and, rolling from the ledge, fall over the precipice and are dashed to pieces on the rocks below. Even the old birds are at times spiteful to youngsters that intrude on their quarters or are imagined to threaten their own offspring. Any unfortunate that has slipped from its nest to some lower ledge during the absence of the parent receives unmerciful stabs from the powerful beak of every adult it approaches, and in the end is either hammered to death or forced over the cliffs. On the north and east sides of the Rock the tide bears away all signs of such accidents. There is, however, on the north-west a ledge a short distance above high-water mark that stretches for about one hundred yards below the cliffs, and over which the waves only break during heavy gales. This ridge is seldom visited; and the sight that meets the view when the upper part is reached is certainly cruel and revolting. Dead and dying Gannets in various stages, from the newly-hatched and naked squab to the adult in perfect plumage, were lying either jammed among the crevices or maimed and mangled in the pools of rain-water when I last made my way to the spot. Numbers of carcasses partly decomposed were observed, and it needed over a dozen shots to put those that had survived the injuries caused by their fall out of their misery. The larger juveniles probably owed their misfortunes to an irritable and querulous disposition, which had brought about contentions with their neighbours; the adults, however (of whom I counted three or four), must have been struck by falling stones. Pieces of rock are not unfrequently dislodged during rough weather, and, dashing down among the crowds of birds, scatter death and destruction on all sides.

The landlord of the inn at Canty Bay, who also hires the Bass Rock, depends mainly on the Geese * for paying his rent. The average take for each season was, at the time of my last visit in 1874, about eight or nine hundred full-fledged young birds. The number, however, depends greatly on the weather, as should the rocks remain wet and slippery from continued rain, it renders the work of going over to collect them both dangerous and unpleasant. To add to the discomforts of the undertaking, the liquid guano not unfrequently lies in pools at least a foot deep on some of the ledges and in the cracks of the rocks. Strong winds also greatly interfere with the successful working of the ropes. Up till twenty or five and twenty years ago as many as 1500 or 2000 were occasionally taken; but since that time the birds have decreased in number, neglect and defective management being most probably the cause of the falling-off.

After being plucked and cleaned, some are sent to the markets at Birmingham, Manchester, and other large provincial towns, and generally bring in from eightpence to tenpence each. Others are hawked about the country in carts and sold for what they will fetch; while a few hundreds are cooked at Canty Bay, and eagerly bought up by the harvest-labourers of the district for a shilling each. A roast Goose also appears to be a favourite dish with some of the visitors at North Berwick. For my own part, I must confess that the stink of the oil and the cooking at Canty Bay set me for ever against such a greasy and highly-scented delicacy. The fat that comes out of the inside of the birds, when cleaned, is boiled down into oil, and sold for farming-purposes at the price of three or four shillings a gallon. In addition to this, the feathers realize from fifteen to eighteen shillings a stone, about one hundred Geese producing sufficient feathers to weigh a stone; these are used for making beds, but have to go through some powerful baking process in order to remove

* In this district a Gannet is always termed a Goose.

the smell of the guano which clings to them. The small village of Canty Bay, as well as the greater number of the inhabitants, usually become thoroughly impregnated with a somewhat unsavoury perfume shortly after the taking of the young Gannets. The process of preparing the birds, and particularly the operation of boiling the fat down into oil, is highly odoriferous; a three months' residence in the cottage of one of the fishermen during the autumn of 1874 gave me many opportunities of ascertaining this fact. When a large number of birds have been brought in, the labour of removing their feathers occupies a considerable time. So gaunt and grim is the aspect of the six or eight elderly hags employed at the work, as they sit in a circle in the dim light of the ruined and gloomy plucking-shed, their heads wrapped round with cloths, half-smothered in dirt and feathers, that a view of the proceedings almost conveys to an observer the impression that he is gazing at an assemblage of witches.

I have frequently been at the Bass Rock during the autumn in former years, when large numbers of young were taken. On the occasion of my last visit in 1874 the weather was exceedingly stormy and interfered considerably with the work. A few extracts from my notes of that year may possibly supply some information concerning the management of the birds and the means of capture that has not hitherto been recorded. The season commenced on the 5th of August, on which day three dozen were procured. Heavy gales and almost constant rain prevented another attempt till the 18th. Though the wind had dropped and the sea was calm, the cliffs were still so foul with liquid guano (which is supposed to be highly injurious to the ropes) that no descent was made, and the work was confined to taking a few young from the upper ledges near the summit of the Rock. For this purpose a long pole with a hook is used, the bird being seized by the neck and drawn to the top, where it is immediately killed. When all that can be reached from one ledge have been procured, they are collected into a heap and thrown over the cliffs into the water; the carcasses are then picked up by a boat which is waiting on purpose to collect them. The younger and more active of the crews from the small village at Canty Bay are generally employed in the work of capturing and killing the birds, while the older men remain in a boat a short distance clear of the Rock, and on some prearranged signal row in and pick up the bodies. John Kelly and Andrew M'Lean, two of the oldest hands (both over 70 years of age), told me they met with a narrow escape, while so employed, a few years back. They had, as usual, been watching at some distance outside to see the birds flung over. After fifty or sixty had been thrown down, as a sudden stoppage occurred, they imagined the men had completed their work, so, pulling close in to the Rock, commenced at once to pick up. It turned out, however, that those on the top had merely rested for a few minutes, having scarcely cast over half the number they had killed. But two or three had been got on board (John Kelly was paddling with the oars, while M'Lean was busily engaged in lifting in the birds with the boat-hook) when crash came a young Goose, striking an oar and breaking it in two pieces. The next moment another was dashed against the gunwale, and a third just missed the head of old John, causing him to let go his unbroken oar. The poor fellows were now perfectly helpless immediately below the spot from which the Geese were being flung. The Rock is so steep that there was not the slightest chance of their being seen from the top, while the clamour of the Gannets (rendered doubly noisy by their recent bereavements), together with the roar of the sea, precluded all hopes that their cries could be heard. Luckily the boat, though old, was strongly put together; and, creeping at once beneath the seats, the terrified crew escaped with no more serious results than a good scare. After having thrown over the whole of the birds they had brought together on the summit of the Rock, the men looked out for the boat, in order to give the signal for collecting the bodies from the water. As the craft remained invisible and a nasty swell was rolling round the Rock, some mishap was anticipated; so hastening down to the landing-place, where their own boat was moored, they pulled round towards the quarter in which the old men had last been seen. At length they were found, little or none the worse, though half-smothered with blood and feathers. Many of the birds which struck the boat had been dashed to pieces, sending the fragments flying in all directions. A young Gannet when just ready to leave the nest is considerably heavier than an old bird, and, falling from such a height, must have struck the boat with immense force.

Some young Geese that had entirely lost the down were taken on the 18th of August for specimens; and I also obtained one in the stage termed by the natives a "Parliamentary Goose." It occasionally happens that the whole of the down disappears with the exception of a thick covering on the head and neck; this much resembles a wig, and gives the bird a most comical expression. I have seldom seen one showing that stage so perfectly as this specimen, whose portrait is given in the Plate. As usual, I remarked many of the youngsters very ill-tempered, snapping and quarrelling continually amongst themselves. On the 19th, wind west and weather fine, we were again out to the Rock, and between three and four dozen young Geese were taken; I also procured four young birds for rearing in confinement. One hundred and eighty-five young Geese were obtained on the 20th. The weather was dull and damp and the rocks were extremely dirty; the ropes, however, were used, though the work was far from agreeable. We were out again to the Bass on the 24th, and five dozen young Geese were brought back. This was the last day of the season I assisted at the operation. Having now gained all the knowledge concerning the "farming of Gannets" that could possibly be acquired in this district, I turned my attention to the large flocks of Skuas which had lately shown themselves in the Firth. As the crew from Canty Bay sailed with me, the taking of the Geese was carried on by fresh hands from North Berwick. The weather, however, was much against a good "harvest," and but few more birds were brought ashore.

On the 30th of September, the wind being light from the south and a calm sea, some large flocks of Gannets were observed about halfway between the Bass and the May Island; these birds appeared to have found fish of some kind (probably herrings) very plentiful and close to the surface. The graceful manner in which the Gannet takes its prey, steadying itself for a moment in the air and then darting headlong beneath the waves, must be seen to be thoroughly understood. Many of the birds were busily engaged dashing down in the usual fashion from a height of twenty or thirty, or perhaps forty, yards; I also noticed several were rising only a few feet from the water, and, after making two or three strokes with their wings, immediately plunging down again, their line of flight in such cases representing a perfect curve of ten or fifteen feet in height and about twenty or thirty yards in length. I imagine, when this method of capturing their prey is followed, that the shoals of fish must be close to the surface. It is only in the autumn and off the Scotch coast that I have witnessed this style of fishing. As far as I have been able to ascertain from constant personal observation all round our coasts, the Gannet (when in a state of nature) does not attempt to dive while swimming. As I have repeatedly watched my tame birds indulging in this performance, it is evident they are quite capable of obtaining their food in this manner, if so inclined. It is, I believe, seldom that these birds prey upon the lythe or pollock whiting, which is excessively abundant in this part of the Firth; I am unable to recall a single instance where I have seen a fish of this species brought as food to the Rock. On one occasion an immature Goose of about three years of age was secured after darting down and seizing a lythe struggling on a line. The Gannet appears to be a good judge of fish, and usually selects only the best flavoured and most nutritious of the finny tribe. Towards the end of the month there were large shoals of mackerel along the shore, and the fish attracted numbers of Gannets; the birds occasionally swept down so close to the rocks, that it was a wonder how they escaped destruction in the shallow water.

Heavy gales occurred early in October; but as the weather moderated on the 7th, I made an attempt to visit the Bass, in order to learn if all the young Geese had left their breeding-places. On approaching the south side we discovered that it was utterly impossible to effect a landing without considerable risk, as a terrible swell was surging round the Rock. In order to obtain a view, sail was lowered and the boat pulled round; it was, however, necessary to keep at a respectful distance, owing to the surf. Though all the breeding-stations were carefully examined through the glasses, not a single young Goose could be distinguished on any of the ledges, the strong wind of the previous week having probably blown off those that had not already taken their leave. A few old birds were flying round and occasionally settling, but the main body had entirely disappeared from this part of the coast.

The natives declare that the young Geese are driven off by the old when able to fly. It is possible that this occasionally happens (and I should imagine it highly probable, owing to the behaviour of my tame birds), though I have repeatedly seen the young leaving the Rock of their own accord. When almost full-feathered they may frequently be observed flapping their wings; and while so engaged I have noticed them, especially in squally weather, lose their balance and come fluttering down from their ledges. In such cases they usually manage to avoid striking the lower part of the Rock and reach the water in safety. On leaving the nest for their first flight, young birds seldom succeed in making their way further than one hundred and fifty or two hundred yards from the Rock, so weighty is their condition that the wings have not sufficient power to enable them to gain a longer distance. The juveniles are supposed to be entirely deserted by their parents when once they reach the water. I have seldom, if ever, met with old birds and young in the first plumage in company at sea; and in not a single instance did I ever observe an adult paying the slightest attention to a young one after leaving the nest. It is a curious fact that not the least notice is taken of a young one falling on the water, even by its own parents; while an old bird that is shot will immediately draw scores around it, where they will remain flying in circles till the bird has drifted a mile or two on the tide.

The men have an idea that the young after leaving the Rock are obliged to remain for at least a week on the water, till they have become light enough to get on wing and procure food for themselves. When a gale from the north-east comes on shortly after a number of young have left the Rock, they are all blown ashore, being unable to make headway against the force of the wind and sea. From my own observations I was perfectly convinced that when they first reach the water after leaving the nest they were utterly incapable of rising on wing. I have rowed up alongside of numbers, and, though driving and even touching them with the oars, they made no attempt to seek safety by flight; it was perfectly obvious that swimming was their only resource for a time. I am aware that young birds when old enough to leave the nest are so well nourished that they are capable of sustaining life for considerably over a week without food. A farmer in the neighbourhood of Canty Bay having requested the men to procure him a young Goose alive, one was brought ashore and placed under an inverted washing-tub till a chance occurred to forward it to its destination. The unfortunate bird, however, was entirely forgotten; and it was not till the fourteenth day that, the tub being required, the captive was discovered. The poor creature was certainly rather light and weak, though by no means in a hopeless condition. The weather, unluckily, happening to be rough, there was no fish at hand, so the wretched bird was knocked on the head to prevent its dying of hunger.

I believe the young separate entirely from the old birds immediately after leaving the Rock. As far as I have been able to judge, they make their way south at an earlier date than the birds in the more advanced stages. On the 23rd of September, 1880, I shot a young one in the nestling-plumage about half a mile at sea, off the coast of Sussex*. I have never met with adults in this part of the Channel till some months later. The Gannets most commonly observed attending the fleets of herring-boats in the North Sea during October and November are the adults and the intermediate stages of two and three years of age. From the 15th to the 20th of December, 1878, large flocks of Gannets were observed in the Channel between ten and twelve miles off the land, and I particularly remarked that not a single bird of the year could be detected; the majority were in the adult stage, though a few exhibiting a certain amount of black feathers on the back (probably between two and three years of age) were occasionally seen. Immense shoals of sprats were doubtless the cause

* This useless murder would not have been perpetrated had I not thought it possible, when the bird appeared in view, that it might be a juvenile escaped from the enclosure where my tame Gannets were confined. A young one had been noticed that morning flapping its wings in a manner that suggested a desire to be off.

of this gathering. Gulls of various species, Divers, Guillemots, and Razorbills were also in attendance in tens of thousands, the surface of the water for miles being perfectly alive with mixed swarms of noisy and ravenous sea-fowl. The white plumage of the countless multitudes of Gannets, as they flew in circles and dashed down incessantly below the waves, appeared, when viewed at a distance against the dark and wintry sky, like clouds of sleet or snow drifting before the wind. On the 13th of May, 1880, I observed two or three immature birds of one year old flying in an easterly direction in the Channel, about five or six miles off the coast.

During the summer months but small numbers of birds of the previous year make their appearance in the neighbourhood of the Bass. In no case are they permitted to intrude themselves on the ledges frequented by the adults; they may occasionally be seen sitting on some of the grassy slopes near the summit of the Rock, in company with the non-breeding birds of two, three, or four years of age. I am of opinion that but a small percentage of the Gannets in the immature stages show themselves at the Rock; most probably the greater number pass their time entirely at sea till they have reached maturity.

The actions of the Gannets on land are somewhat clumsy; their nests, however, are in most instances placed close to the edge of the cliffs, and but few steps are needed to cross the intervening space. When launching themselves into the air, they appear for some distance to gain but little assistance from their wings, a considerable drop being made before they strike out from the rock. The graceful curve effected as the birds swoop out from their breeding-ledges is sure to attract the notice of strangers when viewing their haunts for the first time: on rising from the water they also flap for several yards over the surface, until they attain sufficient power to mount into the air. I much doubt whether a Gannet could rise from the ground, if flat, unless assisted by a strong wind.

Previous to the passing of the Sea-Bird Act, the Geese on the Bass occasionally suffered from the depredations committed by boatloads of strangers, who sailed round the Rock and expended a quantity of ammunition in blazing at the busy swarms engaged with their nests. Numbers were also shot by the fishermen from Dunbar and other parts of the coast. There was some excuse for these poor fellows, as they made use of those they obtained for food, and when bait was scarce a skinned carcass was not unfrequently employed as bait for their crab-pots.

The note of the Gannet is powerful, though far from musical. If interfered with while sitting on their nests (and they seldom make a move unless threatened), the old birds will strike at the aggressor with their sharp-pointed bills, giving vent at the same time to a succession of hoarse, croaking sounds. Before daybreak I have on two or three occasions climbed to the summit of the Bass, and looked down on the silent multitudes collected on the ledges, while the first rays of the rising sun lit up the scene. In almost every instance the male and female were sitting side by side on the nest, the young, if small, being hidden from view, and those of larger size in most instances snugly nestled between the parents. As the daylight increases, first one and then another stretch out their necks and, uttering a low note, rise up and flap their wings. It is soon an animated sight: the old birds may be seen on all sides rubbing their heads together and going through the most amusing antics, the larger nestlings frequently thrusting up their heads between the pair and joining in the performance. When once the day's work has fairly commenced there is a constant clamour from all quarters, as the birds in rapid succession start off in search of prey. The cry of the young in the first instance is a feeble squeak, which shortly increases in strength, but does not attain full power till after leaving the nest.

Some remarkable descriptions of Gannets' nests have appeared in print, in which the structures were stated to have been piled up to an altitude of several feet. I am at a loss to account for such assertions, having seldom noticed one where the materials were accumulated to a greater height than six or eight inches. Seaweed and tussocks of coarse grass, torn from the pasturage near the summit of the rock, are used in building, fresh supplies being continually added, even after the young are hatched. The birds frequently steal from one another, and an unguarded nest is seldom left long without being systematically and deliberately plundered by the neighbouring pairs. Numbers of birds may be noticed at all times during the summer flying towards the Rock, bearing in their beaks large lumps of seaweed, which they have secured floating on the water. The materials made use of soon become compressed, owing to the constant weight of one or other of the parents, and decompose so rapidly from the effects of the guano, that there is no chance of the quantities of rubbish collected ever assuming any excessive bulk.

The egg is singular in texture and appearance. When fresh-laid it is of a clear white or bluish tint, coated here and there with a covering of a chalky nature; the shell rapidly assumes a darker hue, stained by the dirt from the nest and discoloured by the filth that clings to the webs of the parents as they waddle over the slimy ledges.

The aspect of the Bass Rock during summer and winter differs considerably. After the departure of the birds, the conspicuous stains of the guano rapidly disappear from the face of the cliffs when exposed to the force of the winter gales, and a transformation from white to a dark grey at once takes place.

It may not be out of place to give a list of the birds that usually breed on the Bass. The numbers of many species have considerably diminished during the last few years, and others, formerly abundant when I was first acquainted with the Rock, had at the time of my last visit almost entirely ceased to put in an appearance during the nesting-season.

Peregrine. A pair not unfrequently attempt to rear their young on the Rock. I am afraid their efforts are seldom successful, having noticed them on several occasions making use of very accessible ledges.

Kestrel. This species is not a regular breeder; but I have repeatedly seen a pair circling round the rocks, and on one occasion obtained a view of the young on a ledge in the deep crevice on the west side.

Jackdaw. These robbers were numerous between twenty and thirty years back, and, having made themselves excessively disagreeable by preying on eggs, excited the wrath of the tenant, who thinned them down by poison. They formerly nested in the rabbit-burrows near the summit and also in parts of the buildings. At the time of my last visit they were exceedingly scarce. Though a few were reported as breeding, they escaped my notice.

Rock-Pipit. This Pipit nests in some numbers on the Bass, principally among the old ruins, but in a few instances on the ledges on the south side.

Blackbird. A few pairs of these birds are during most seasons to be found among the buildings. The nests are usually placed among the ruins of the old fortifications; and I also discovered two or three during different seasons in a sheltered nook on the west side. A situation was chosen on one occasion within a few yards of the eyrie of the Peregrine.

Sheldrake. I saw one clutch of eggs taken off the Rock. The nest was placed in a rabbit-hole in a small open piece of ground amongst the lower part of the buildings on the south side.

Eider. A pair used now and then to lay in sheltered corners among the lower range of buildings. The treatment they met with was not encouraging; and after being robbed for several successive seasons, the birds entirely deserted the spot.

Gannet. By far the most numerous of the summer visitors.

Common Guillemot. Numbers resort to the Rock; but as they mostly frequent the higher ledges, a few only can be detected from the water. The Ringed Guillemot may be seen in small numbers every season.

Razorbill. Not nearly so plentiful as the Guillemot. These birds have decreased greatly during the past few years. Their favourite positions appear to be the higher ledges on the north-west side.

Puffin. Far more abundant than a stranger would imagine on paying a mid-day visit to the Rock. The holes in the masonry of the old fortifications are their quarters.

Shag. A pair occasionally nest in the east cave; and some years back I saw them frequenting the west side.

Lesser Black-backed Gull. Seldom more than a pair or two now seen resorting to the Rock during the breeding-season. The nests are placed on the grassy slopes near the summit.

Herring-Gull. Scarcely more plentiful than the last species, though I have discovered as many as three nests in one day while searching the ledges among the detached slabs of stone on the higher parts of the Rock. Many pairs of this Gull and also of the Lesser Black-backed fell victims to the poison laid out for the Jackdaws; they have never regained their former numbers.

Kittiwake. The most numerous of the Gulls. There are hundreds of nests on the narrow ledges in the steep face of the Rock on the north, east, and west sides.

The above list is made up entirely from my own observations on the Rock between 1862 and 1874. I have not visited the locality since the latter year, and prefer to trust only to my own experience. I am aware that the Cormorant, Great Black-backed Gull, and Common Gull are reported to have bred on the Rock of late years: if such has been the case, I can only state that their nests entirely escaped my notice.

A few remarks extracted from my notes concerning the Gannets brought from the Bass and reared in confinement may supply some information which could scarcely be obtained while studying their habits in a wild state.

"1874, August 10th. Four young Gannets taken from the Rock. These birds were full-fledged and proved exceedingly troublesome to feed.

"September 10th. The captive Geese were now more tractable, and took their food when offered to them without needing to have it rammed down their throats. I imagine the whiting and haddies with which we were obliged to feed them were not so acceptable as the herrings or mackerel usually supplied by their parents; it was, however, impossible to procure other fish at Canty Bay. The amount of phosphoric matter in the whiting is very remarkable. Often when passing at night the sheds where the birds were kept, I noticed that any fragments of fish left from the previous day, and even the boards on which they had been cut up for the small birds (Guillemots), gave out a pale luminous vapour, which seemed to hover round the spot, flickering up more brightly from time to time.

"1875, August. Early in the month received two young Geese from Canty Bay. These were younger by at least a fortnight than those taken the previous year, and very shortly became reconciled to confinement.

"1876, August. During the first week two young Geese were sent from Canty Bay. Profiting by the experience of former years, birds which were but half-fledged had been selected, and far less trouble was caused."

The various changes of plumage exhibited by the Gannets in confinement during the immature stages can readily be traced by an examination of the Plates.

"1879, March. Early in the month a nest was built in the shed to which the birds resorted for shelter during the winter months. The structure corresponded precisely with those I have examined on the Bass, being composed of seaweed and coarse grass. These materials had been supplied for some weeks, the pair having previously attracted attention by collecting sticks and feathers, as well as tearing up the grass round their enclosure. It was seldom that the nest was deserted by both birds, one or other being almost always on guard.

"May 4th. The female had continued on the nest for some days, and it was at last discovered she was covering an egg. After sitting five or six days the younger birds succeeded in dragging away some of the materials and smashed the egg. The nest was reconstructed immediately; but though the bird continued sitting for several days, it was eventually deserted.

"1880, February. Materials for building were collected by the old pair during the month, and a foundation of the nest was laid down. Constant encounters with their neighbours, who persisted in stealing whatever they could lay hold of, somewhat delayed the completion; and it was not till well on in April that the structure assumed anything approaching the correct proportions. From this date it was seldom left unprotected.

"May 19th. An egg was noticed this morning in the Gannets' nest.

"June 4th. The pair continued sitting closely, the nest being at all times tenanted by either the male or female. One or the other was almost incessantly employed in carrying fresh materials (grass, seaweed, and feathers) and adding to their nest; the structure, however, being flattened down by their weight, increased but little in size or height.

"20th. On visiting the Gannets in the evening, I remarked the male and female both asleep by the side of the pond and the egg unguarded. This was the first time it had been left exposed to view.

"21st. The female was again sitting. For several days the other pairs of Gannets were uneasy, and frequently attacked the sitting bird; possibly they were enraged by the pertinacity with which all their nesting-materials were seized by one or other of the old pair and appropriated.

"30th. The glimpse that was obtained of the egg showed that it was cracked and chipped.

"July 1st. In the morning the young bird (a small, black, shapeless monstrosity resembling a toad) was partly clear of the shell, and fully hatched by mid-day. The female had apparently not left the nest for several days.

"2nd. The male Gannet mounted guard on the nest, and both remained side by side till night. I noticed the young one stretching its small naked wing-joints while the parents were shifting positions over it.

"3rd. One of the old birds plucked a number of feathers from its back, placing them carefully on the nest. This was a somewhat useless proceeding, as on flapping its wings shortly after the whole were blown away.

"7th. Both old birds sitting on the nest. The young one now showed the eyes open: iris dark hazel; eyelid and circle livid slate-colour; beak a dull slate, white at the point; a slight sprinkle of down on the head. The body was invisible under the old birds.

"8th. Obtained a better view of the youngster. There was no down except on the head. Both parents on nest almost continually.

"9th. Young Gannet very noisy, squalling in the most vigorous manner. The old birds were feeding it on several occasions when visited.

"10th. The down on the nestling increased on the head and sprouting all over the back. The beak was now about three quarters of an inch long.

"13th. This was the first day on which I obtained a chance of closely watching the operation of feeding the young bird. Had I not previously viewed the performance at a distance on the Bass, I should certainly have imagined the young one was about to be swallowed by an unnatural parent. The old bird, for a few minutes before commencing, gave evidence, by certain movements in the neck, that the food was being prepared and gradually brought up. The nestling was calling faintly and lifting up its head open-mouthed, when the old bird dropped forward, and, opening the beak to an enormous extent with the head drawn sideways, apparently scooped the young one into its mouth. Being almost entirely concealed, the actions of the young one could only be conjectured. From having frequently observed their movements when of larger growth, it is evident that they help themselves to the preparation ejected into the throat of the parent.

"18th. Young Gannet much increased in size, but no more down showing, less, in fact, being discerned than was noticed some days previously. Doubtless the grimy feet of the parents, almost perpetually spread over the back, were responsible for the apparent diminution.

"19th. Male and female fighting most viciously, the male apparently desiring to come to the nest and the female refusing to leave. During the struggle the male was almost choked. After having been seized by the throat by his spouse, who resolutely refused to let go, he dragged her from the nest in his attempt to escape, and the unfortunate babby was precipitated out of the nest *. As the accident was witnessed, and the young one immediately returned to its quarters, no harm resulted.

"20th. Young bird much increased in size and down.

"24th. Male and female Gannets once more friendly, and sitting cosily side by side on the nest. Young bird now increasing rapidly. The down by this time was thicker on the body (where it had been so long in making its appearance) than on the head and neck. The old birds now feed the young one, standing over and opening the mouth (the jaws seeming in some manner to unhinge to an unnatural extent), the young one rising up and stretching well into the gullet.

"28th. The old birds did not now stretch down their heads so far when feeding. The food supplied consisted of good-sized pieces of mackerel, one third of a fish being noticed to fall down the young one's throat.

"31st. While feeding the young one the old male dropped a whole mackerel from his throat, but immediately snatched it up and reswallowed it, duly correcting the babby, who had made an attempt to appropriate the fish. The young bird almost entirely covered with down up to the black face.

"August 9th. The manners of the Gannets are not nice. Both old birds were sitting on the nest while the male was feeding the infant. A herring and several pieces happened to drop from his throat, when a portion was seized by the young one, while the female at once laid hold of the herring, leaving only the remaining bits to her lord and master, who forthwith gobbled them up and then waddled off to the pond †.

"11th. Young Gannet apparently fully covered with down, but no feathers as yet showing.

"18th. Tail-feathers just commencing to show on the young bird.

"21st. The tail was now plainly visible all round, and the wing-feathers were just appearing.

"26th. The feathers were now coming thickly on the back and wings.

"Sept. 1st. Young Gannet getting well fledged on back and wings; feathers also showing plainly on the forehead; this part being bare black skin, the feathers resemble small white specks.

"6th. The young bird now took food when offered, and succeeded in swallowing a couple of mackerel placed in its mouth.

"13th. Nearly all the down disappeared off the back of the young bird, the feathers showing through the down all over the breast.

"17th. The young Gannet was today left in the nest for a short time, unattended by either male or female. Very little down now visible: a small bunch on back of neck and under the throat, and a few patches on the flanks.

"19th. Young Gannet attacked severely by the male, who seized the poor bird by the neck and, forcing it from the nest, dashed its head against the woodwork of the shed. In order to prevent a recurrence of this treatment, the nest was shut off with wire-work, so as to keep the old birds at a distance.

"20th. As the young bird appeared uneasy in confinement, continually attempting to get through the wires, and causing the beak and feet to bleed, it was released. The parents for some time would take but little notice of their offspring, though the poor bird at once made its way towards them and cried for food ‡; at length one and then the other would occasionally pluck a little of the down from the head.

* Dead youngsters round the nests and on the ledges on the Bass plainly show the rough treatment they go through when their parents fall out.

† These facts are recorded somewhat minutely, as they assist in accounting for the quantities of decomposing fish noticed on the breeding-ledges. Where the birds are thick and frequently snapping at one another, as well as dragging away portions of the nests, it will be readily understood how the fragments collect.

‡ The natives of Canty Bay (who ought to be well acquainted with the habits of the Geese) declare that the old birds take no notice of the young after leaving the nest. According to my own observations, this statement appears correct.

"21st. The old birds again took up with the young one on the opposite side of their enclosure, both male and female attending to it.

"22nd. Old birds still with the young one, and occasionally driving off any of the other birds that approached too close for their liking. They still supplied it with food, sometimes in the form of soup and at times whole fish.

"24th. The male had done the greater part of the feeding, and I again observed him bringing up portions of fish, which the young one took by thrusting its head down the parent's throat.

"25th. Having noticed the young Gannet continually spreading its wings, I had them clipped today. I did not imagine it could rise under ordinary circumstances from so small a space; it was, however, quite possible the bird might be carried away during a sudden gust of wind. The old birds were busily employed in plucking off the small portion of down that remained on the head.

"27th. The young Gannet had now lost every particle of down, but was still looked after by the old birds.

"October 2nd. This was the last date on which the young one was observed to be fed by the parents.

"1881, March 30th. The young bird had by this time changed but slightly from the nestling-plumage. There were a few white feathers showing about the breast, neck, and back of head. Beak dull pale horn; iris a pale lead-tint; circle round eye pale lead, with a bluish tinge; feet black; the markings on legs and toes an indistinct dirty white."

The same pair of Gannets nested again in 1881, the egg being laid on May 8th, the young one hatched June 21st, and full-fledged by the 4th of September. It would be useless to record a second time the observations concerning the development of the plumage of the nestling: in almost every particular its progress corresponded precisely with the growth of the young one of the previous year. One fact, however, ought not to be omitted, as it tends to bear out the impressions I had formed on the Bass, viz. that the adults will not tolerate the presence of birds in the immature stages in the vicinity of their quarters. On the 11th of July, while the male and female were engaged with the latest baby, the other four adults (whose constant squabbles over their own nesting-arrangements had put a stop to all hopes of domestic felicity) suddenly, and without the slightest cause, set upon the unoffending young bird (now just turned a twelvemonth old), and so severely injured it that recovery was hopeless.

I have seen statements to the effect that the Gannet is unsuited to confinement, and ill repays the consideration with which it is treated. The poor creatures are by nature endowed with a voracious appetite, and, if starved, necessarily become ravenous and possibly spiteful. When looked after by those acquainted with their requirements and willing to supply them with a sufficient quantity of food, none of the feathered tribe could be found whose habits are more interesting, and but few so harmless and gentle.

From the time they were first removed from the nest, the young birds I procured at the Bass had varied in temper and disposition. Some were particularly confiding, following those they were acquainted with round their enclosure, endeavouring to draw their attention; while two or three were morose and shy, resenting the slightest familiarity with harsh screams and a lunge from their powerful bill. With a single exception, they became more amiable as they progressed in years. While in the first plumage their appetite is far greater than in the more advanced stages, six, eight, or even ten herrings being occasionally swallowed at a single meal. I have repeatedly watched a bird with the tails of three or four herrings protruding from its mouth, and still eagerly looking out for another fish. Two, or even three, of the largest Irish mackerel were at times consumed by one individual, though the bird for some minutes exhibited signs of extreme discomfort, being incapable of bending its neck or even turning its head. The number of sprats that a hungry Gannet could consume would, I should imagine, amount to several hundreds. Mackerel, herrings, and sprats appear to be their favourite food. It is strange that they would at once reject a pilchard if offered to them amongst the herrings, dropping the fish the moment they had seized it. When herrings were not to be procured, I have known them accept the pilchards as a last resource; but they were evidently taken without relish. I tried them with several other varieties of the finny tribe, such as pollack, haddock, whiting, mullet, and smelts, but not one would they ever swallow of their own accord. During hot sultry weather in July and August I often remarked the birds were but little inclined for food. A fish would occasionally be taken, but after being tossed in a sportive manner from one to the other, it was usually relinquished. For a week, at times, this behaviour was noticed, though now and then they might be tempted to accept a fish or two by offering them their food towards evening, when the fierce heat of the day had passed.

The walking-powers of the Gannet can seldom be studied while the bird is in a wild state. It is usually allowed that they are extremely awkward: this is certainly the case on rough ground, but over an even surface they are able to make their way with about the agility of a farmyard Goose. I soon learned that the great difficulty (indeed the only one) in keeping the birds in health was to provide them with a suitable exercising-ground. In the first instance, the enclosure in which they were turned out was laid down with a foundation of chalk about a couple of feet in thickness, and on this fine gravel was spread. Two or three of the birds soon showed signs of lameness, and the soles of their feet became swelled. As I considered the gravel too rough it was removed, and sea-sand put down in its place to a thickness of about six inches. For a month or so this answered admirably, and their feet (the swellings having previously discharged) rapidly recovered. I had just come to the conclusion that the sand was all that could be desired, when the weather, which had for some time been wet and dull, rapidly changed, and after a few hot days a strong breeze, increasing into a heavy gale, set in. The next time I visited the birds the aspect of their enclosure was completely altered; the whole of the sand that had not been blown into the pond was banked up like a snow-drift along one side. As the place was much exposed, and consequently always liable to be affected in this manner by the force of the wind, the sand was removed, and the surface of the chalk rolled and beaten down as far as possible. This appeared to suit the birds well; with a single exception, they had no return of the swellings, and the one that suffered occasionally, recovered rapidly when once the corn broke and discharged. After the chalk had been down for a year or two, small patches of grass began to show themselves. I then had grass-seed sown thickly, and the place constantly watered; it soon became completely covered with a fine short turf, which, owing to the chalk foundation, never became damp or spongy.

The young birds at the age of five or six months, with the exception previously referred to, showed themselves as playful and mischievous as a litter of puppies. They proved, however, somewhat destructive to plants and shrubs: any branch or stick they were able to tear up was immediately seized hold of and dragged from one to the other for hours. A Water-Rail, which had been caught and turned into the garden where they lived, was a source of great amusement to the Geese. The active little bird appeared but slightly disconcerted by the treatment; at least, he was perpetually trying how near he could get to them without being snapped up. On one occasion, having incautiously ventured too close to a Gannet that was merely feigning sleep, he was seized in a moment. For a second or two the unfortunate bird disappeared in the capacious throat of his captor; but before I could come to his rescue was again free, and, aided by legs and wings, speedily gained the friendly shelter of the bushes before the other Gannets which were waddling up could cut off his retreat.

That a wild Gannet would molest any diminutive member of the feathered tribe is improbable, though the tame birds not unfrequently amused themselves in this manner. I happened to be watching a party of Sparrows making an attack on the corn provided for some tame Gulls, when a Gannet, who was resting with his head drawn back and apparently perfectly unconcerned with what was going on, suddenly dashed out his neck, and seizing an unlucky Sparrow, bolted it in a moment. Considering the skill with which the seizure was effected deserved to be rewarded, I hit upon a plan to render the capture more easy. Removing the board on which the corn was spread, some round holes were cut, in which were inserted neat little paper cones filled halfway up with corn and smeared round the edges with bird-lime; then returning the board to its usual place, I waited to see the result. The Gannets appeared thoroughly to understand the whole proceeding, and never interfered with the Sparrows till they got into difficulties; but as one cunning old bird succeeded in swallowing three Sparrows that were fluttering about, together with the paper cones, bird-lime, and corn, I thought such a mixture could scarcely be beneficial to their health, and accordingly gave up the experiment.

The Guillemots kept in the same enclosure were also fond of a quiet joke in their own small way. Whenever they could catch a Gannet asleep on the bank with its tail hanging over the pond, they would swim close up, and, seizing the feathers in their beak, give two or three good tugs, and then disappear below the surface before the astonished Gannet could make out from where the attack was made. They had, however, to be careful that they were not caught, as what was play to a Gannet was death to a Guillemot, and once or twice they paid for their fun with their lives. On one occasion a poor bird was confused by two Gannets snapping at him at once; and one seizing him by the head and the other by the feet, he was dead before an attendant, who had watched the affair, could get to the rescue. Although appearing almost impossible, it is doubtless a fact that one was entirely swallowed by a Gannet. For several hours a Guillemot was missing, and, owing to a strong breeze, it was conjectured the bird must have got on wing (they were never pinioned) and flown away; the following morning, however, it was floating on the water in the pond in a condition that left no doubt as to its fate. The feathers and flesh had entirely disappeared from the head and neck, and the skull was bare; the colour of the beak was also faded to a livid flesh-tint. It was then remembered that one of the Gannets had refused his dinner on the previous day; and, considering the lunch he had made, this could hardly be wondered at. When it is considered that these birds can swallow at one time two or three of the largest mackerel, it is not so surprising that the feat was accomplished, though I am unable to account for the inducement.

I never remarked the Gannet in a wild state make an attempt to dive for prey while swimming. Those I kept in confinement, having been pinioned, were unable to rise on wing; they were, however, by no means incapacitated from reaching any fish that sunk to the bottom of their pond. The whole party might frequently be noticed busily engaged in diving; their actions (which certainly lacked the ease and silence with which Cormorants or Divers disappear beneath the waves) closely resembled the plunge that is made by the Coot. The wings were used below the surface, after the manner of the Guillemot.

When Gannets make the downward plunge straight from the air (previous to which they occasionally sail round and steady themselves for a moment), it is probable that they have detected the presence of fish and selected their victim beneath the surface of the water. Their sight is excessively keen; and the accuracy with which the tame birds would catch fish when flung to them from a distance of twenty or thirty yards was certainly surprising. If thrown within a yard or two of where the bird was either standing or swimming, not one fish in twenty would be dropped. In case the prey happened to be seized in some manner that was unsuitable, it was usually tossed up in the air to the height of perhaps a foot, then taken head downwards and immediately swallowed.

The specimens from which the figures in the Plates are taken were all wild birds, the ages of those exhibiting the plumage of one, two, three, and four years being judged by comparison with birds reared in confinement. The ages of the nestlings in the more advanced stages were also ascertained in the same manner, the dates on which the younger birds hatched out on the Bass being observed and noted down.

In Plate I. are figured an adult female and a nestling between five and six weeks old.

The principal figure in Plate II. is a nestling between eight and nine weeks old. The usual and demonstrative greeting that takes place between an adult male and female, on the return of one of the pair to their nest after a lengthened voyage in search of prey, is depicted in the background.

Plate III. gives the full-fledged young bird in the state in which it leaves the Rock, probably twelve or thirteen weeks old. The stage termed by the natives the "Parliamentary Goose" is also shown, the fringe of down round the head and neck being supposed to bear a resemblance to a wig. The youngster was from ten to eleven weeks old; and I have seldom remarked the downy wig so perfect as in this specimen.

The two Geese in Plate IV. are just over one year old. The difference in the white markings may be accounted for by the fact of one having been hatched somewhat earlier in the season. In a few weeks' time the darker specimen would be in much the same state.

In Plate V. the same reason may be given for the slight difference in the plumage of the specimens. These birds were but little over two years old.

A three-year-old Gannet and the last stage (four years old) before assuming the perfect adult plumage are shown in Plate VI.

SANDWICH TERN.

STERNA CANTIACA.

During the seasons of migration in spring and autumn this Tern is to be seen along the coast-line in many parts of the British Islands. In Sussex, Kent, Norfolk, and along the shores of the Firth of Forth, I have repeatedly met with this species working north in spring. On one occasion a party of five or six were recognized in June on the Dornoch Firth off Tain; though the birds remained fishing up and down the channels at low tide for a couple of days, they eventually disappeared, and I was unable to learn that any bred in the district. I often remarked that the birds seen so late in the season exhibited a dark marking on the shoulder of the wing, somewhat similar though less plainly defined than on the immature in autumn. It is probable that, like many other sea-fowl, this species does not attain maturity at the age of one year; doubtless the stragglers are non-breeders, simply passing the summer in making a certain migratory movement towards the north and subsequently working back to their winter-quarters *. On the 2nd of July, 1870 (weather rough and frequent squalls of rain), a pair of Sandwich Terns were observed flying in an easterly direction along the beach at Shoreham. This is a somewhat unusual date for Terns in this locality, and in all probability the birds were immature, though the distance at which they were seen precluded all hopes of carefully examining their plumage.

In autumn Sandwich Terns are to be observed fishing along almost every part of the eastern and southern coasts that I visited during the months of August, September, and October. Stormy weather appears to set these birds in motion; during gales of wind in September and October 1874 they proved unusually abundant in the Firth of Forth, flocks numbering from ten or a dozen up to twenty or thirty being often in view flapping either east or west along the coast. As no breeding-stations were then tenanted in the Firth, it is probable that these birds had worked round the shore from the Fern Islands. I also noticed several small parties just off the parade at Penzance on the morning of the 8th of October, 1880, hovering over the broken water as the swell subsided after the terrific seas that had rolled into Mounts Bay at daybreak †. For a few days after the gale Terns, principally of this species, were making their way along the coast towards the west, the last (a bird with an almost white head, resolutely refusing to approach within range) that came under my notice that season being observed fishing off the harbour at Lamorna Cove on the 5th of November. On the 3rd of September, 1883 (a heavy gale from the south-west), numbers of Sandwich Terns were sheltering in Shoreham harbour, and

* None of the adults procured as specimens during the breeding-season exhibited the dark markings on the shoulder.

† So tremendous was the force of the seas rolling into the bay on that disastrous morning, that the breakers repeatedly dashed over the baths, a building of three stories standing on the parade, and burst into clouds of spray at least thirty feet over the roof. I watched five fishing-boats sailing in from the open sea for the harbour; four succeeded in making their way in, but the last, after passing the head of the south pier, was completely overwhelmed by a tremendous wave that broke over, and the whole crew of seven men were drowned.

several were blown far inland. While driving near Brighton, a couple of miles from the coast, I noticed a single bird trying to make headway against the gusts of wind, which occasionally forced it to fall back: in no other instance have I met with this species at any distance from salt water.

Though several stations where Sandwich Terns reared their young are stated to have existed on various parts of our coasts, the majority appear to have been deserted of late years. It is only on the Fern Islands that I have met with these Terns during the breeding-season; at the time of my last visit, in the summer of 1867, there were probably a couple of hundred pairs. The birds then resorted to two of the smaller islands, where they laid their eggs among the fine gravel and shingle. The men who look after the sea-fowl had surrounded these spots with some large blocks of stone to preserve the eggs from the force of the wind, as they are often destroyed by the terrible gales that not unfrequently break over these rocky islets. One colony was intermixed with Arctic Terns, while the other had selected a more retired stretch of sand partially sheltered by some large slabs of rock; in both instances a few Oyster-Catchers and Ringed Plovers had taken up their quarters near at hand. These Terns lay two eggs in a small depression in the fine gravel, a few strands of grass occasionally collecting; their cradle is honoured by some with the title of a nest. I was informed by the egg-gatherers that they usually change their breeding-places, seldom laying for two successive seasons on the same island.

The state of plumage exhibited by the adult in autumn is shown in the Plate: about the end of June the black feathers on the head gradually become speckled with white, which increases as the season draws on. As to the age of the other bird I am unable to offer an opinion with any degree of certainty, though it is probably in its second year. This specimen was shot on the 10th of September, 1874, flying in an easterly direction in the Firth of Forth, off Canty Bay: the adult was also obtained in the Firth a few days before.

ROSEATE TERN.

STERNA DOUGALLI.

The numbers of this handsome Tern to be seen frequenting our coasts during summer have greatly diminished during the last twenty or thirty years; several of their breeding-stations are entirely deserted, and others where a busy swarm might have been observed during the whole of the nesting-season are now tenanted by but one or two pairs. This species has come under my observation only on two occasions. The first I met with was killed at the Fern Islands early in June 1867: the second flapped slowly past the punt on Breydon mudflats on the 26th of May, 1871; both barrels of my gun having been discharged a moment previously, the bird was out of range before another cartridge could be inserted.

The Roseate Tern shot at the Fern Islands was obtained by the merest chance; two pairs of Arctic Terns had just been secured for specimens, when it was ascertained that the tail-feathers of one of their number had been damaged, and another was needed to take its place. The whole of the birds breeding in the vicinity had been disturbed by the shots and were hovering in a cloud above the boat, rendering it a difficult matter to select a satisfactory specimen and fire without causing unnecessary destruction. At length a Tern with long tail-feathers exceedingly well developed was noticed at a fair distance circling round, apart from the noisy throng, and falling dead on the water without a feather displaced by the charge; the rosy hue on the plumage of the breast at once attracted attention and revealed the species. As the bird proved to be a female and evidently sitting, it was obvious that if no other pairs were breeding on the islands her mate at least must be near at hand; the daylight, however, was drawing to a close by the time our specimens had been packed away, precluding all chances of further search till the following morning. An early start having been effected from North Sunderland, the islands were reached soon after sunrise, and every stretch of shingle or sandy soil to which the Terns resorted was thoroughly explored. Though the occupants of each of the three large breeding-stations as well as the smaller colonies underwent a lengthened examination through the glasses, while the greater number of the birds were either covering their eggs or perched quietly on the adjacent stones, and consequently affording every chance for careful inspection, I failed to detect a single rosy breast. Terns invariably rise on wing should their haunts be approached, and fly with angry screams towards the intruders on their domain; it is by no means an easy matter to identify accurately any single individual in this dense cloud, as darting rapidly or sailing round on expanded pinions each pursues its own course. In order to obtain a clear and uninterrupted view of the whole assemblage at each station after alighting at their nesting-quarters, I made use of the tactics often successfully employed with the Crow family or the larger birds of prey. In company with three or four of the crew of the fishing-craft that had piloted us from the harbour *, we approached one of the colonies, and selecting a spot, at the distance of about sixty yards,

* On a previous visit to North Sunderland, I had remarked that none of the boats in the harbour were suitable for shooting-purposes or following a wounded bird with sufficient speed to ensure a capture, and consequently on this occasion I had forwarded a light 20-foot boat

where rough stones and litter were scattered among the slabs of rock, a shelter that afforded ample concealment was with the help of a piece of old sail-cloth speedily rigged up. After completing the work and placing the finishing touches on my hiding-place, the men withdrew towards the boats. A very few minutes had elapsed when the Terns, after following the disturbers of their peace for some distance, gradually reappeared on the scene, and after hovering round for a time without detecting the alteration that had taken place, the main body settled quietly down, though a few still continued on wing. The greater number of those that had alighted shortly betook themselves to their domestic duties, others were busily occupied in cleaning their plumage, and the remainder, after stretching and going through various contortions, buried their heads in the feathers of the back and sought repose. Ample opportunities for making good use of the glasses were now afforded, and after awaiting further arrivals for over half an hour, I was satisfied that no Roseate Terns had taken up their quarters at this station; then moving on to the next, I was enabled in the course of the day to make a thorough inspection of the whole of the Terns on the islands. The following day was passed in watching the flocks and straggling parties of Terns that kept at sea, either fishing in the bays around the islands or along the sandy shores of the mainland. Although the birds proved exceedingly fearless and were closely approached and examined, I failed to detect in their ranks any of the conspicuous strangers of which we were in search. In all probability but one pair of the Roseate Tern had been breeding this season on the islands, and the death of the female doubtless accounted for the disappearance of her mate from the scene of his bereavement.

Statements have lately appeared in print to the effect that, as a breeding-station, the Fern Islands are now entirely deserted by this species. This may or may not be the case; I possess, however, the best evidence that scarcely a year has passed up to the present date without specimens having been either seen or procured in the immediate vicinity of the islands. The Isle of May, off the northern shores of the Firth of Forth, is enumerated among the breeding-haunts of the Roseate Tern; in this case I am inclined to believe some error has arisen, having frequently visited the spot without observing the birds, or gaining any information that would tend to substantiate the fact.

The tints on the breast of this species, when seen in life or immediately after death, are far deeper and richer than even the most enterprising colourists have ventured to depict; the rosy hue, however, soon commences to fade, and in less than a hour a considerable alteration has taken place. The depth of the colouring doubtless varies considerably in different individuals, and also according to the season of the year. In August 1864 I examined a specimen, shot by a gunner on the east coast, a few minutes after it had been picked up, and remarked that the bird by no means compared in brilliancy with others previously seen. It is, I am of opinion, only through May and the early part of June that the rosy tints are to be seen in their full beauty.

Few opportunities for observing this species during life having fallen to my share, I am unable to supply the slightest information concerning its general habits. The figure in the Plate, which is taken from the specimen obtained at the Fern Islands early in June 1867, renders a description of the plumage unnecessary.

---

of my own from North Berwick. My men being unacquainted with the coast, and also in order to ensure safety should a gale suddenly spring up, we engaged a fishing-lugger to keep in attendance and convey stores in case we should remain on the islands for any length of time.

COMMON TERN.

STERNA FLUVIATILIS.

In spring and autumn, while on the way to and from their breeding-quarters, these Terns may be observed in considerable numbers off various parts of our coast-line; the first-comers usually put in an appearance towards the end of April, and all through May a stream of birds in larger or smaller parties continues at short intervals to pass onward towards the north. The most general movement appears to take place about the middle of May, when immense flocks are occasionally met with in the channel heading steadily on towards the east; after reaching the open sea their course is turned for the north, some making for their breeding-places on the shores of the firths and lochs of the Highlands, while the remainder continue their journey to more distant lands across the ocean.

Many of the breeding-stations to which the Common Tern resorted in former days have been deserted: these birds are stated to have reared their young in considerable numbers on several of the wide-stretching shingle-banks along the coasts of Kent and Sussex; I doubt, however, if a single egg has now been laid on several of their former haunts in this locality for some years. Common and Arctic Terns not unfrequently breed in company: I often passed a mixed colony of these birds on a ridge of low-lying land running down to the Dornoch Firth between Morangie and the Meikle Ferry, to which the name of Ardjackie Point was given. Several of the nests were placed in a field of backward oats, and others among the shingle and rough stones stretching down to the sandy flats. So late as July 2nd, I find in my notes for 1868, there were but few broods of young birds to be seen; in all probability the earlier clutches of eggs had been carried off by the country people living on the hill-side near at hand.

Early in June 1867 I passed several days on the Fern Islands, and closely examined the Terns breeding at all the stations; the only species I was enabled to identify were the Sandwich and Arctic, with the exception of a single Roseate Tern. I was not then aware that the Common Tern had been stated by several authors to nest on the islands; not a single specimen, however, was observed, nor did I meet with the nest of a Tern with more than two eggs for which the rightful owners were responsible. In one instance it was obvious that an Oyster-Catcher had laid in the nest of a Sandwich Tern, and the third egg in the only nest of the Arctic Tern that contained above the accustomed pair had evidently been deposited by a Ringed Plover. From repeated observations concerning the nesting-habits of Terns, I am of opinion that the Common Tern usually lays three, and the Arctic Tern invariably two eggs.

Along the shores of the Channel in East and West Sussex, and on the sands and mudbanks off the coast, as well as up the rivers of Norfolk, are the only places where I have met with opportunities for carefully watching and making any lengthened observations on the spring and autumn migrations of Terns. Common Terns, when on passage, usually fly in large flocks, composed entirely of their own species, though they join at times in company with Arctic and Sandwich. Terns when passing to and from their summer-quarters do not fly in such dense bodies as Plovers, Wildfowl, or Gulls, their ranks being far more open and extended. On several occasions during the month of May, while at sea in the Channel, I have seen uninterrupted streams of Terns passing from west to east without intermission for twenty minutes or half an hour, the whole space in view as far as the eye could reach being scattered over with flocks, small parties, or single birds. It is by no means easy on such occasions to distinguish the Common from the Arctic Tern; the Sandwich, however, the largest and most attractive of the family that regularly pass our shores, may be recognized by the more conspicuous black cap, and the lighter hue of the exquisite silvery grey of the back and wings. The few Black Terns that now approach our islands while on their way to their summer-haunts are also occasionally seen, their darker colouring instantly drawing attention and proclaiming the species as the birds flap past in company with their more brightly tinted relative. Those tiny travellers the Lesser Terns may be at once recognized when fairly in view, their neat and diminutive forms and more jaunty flight rendering all doubts as to their identity impossible.

Under date of May 21st, 1874, I find an entry in my notes relating to the movements of Terns along the south coast:—"Wind south, weather still and fine. Out at sea off Brighton. Thousands of Terns passing along the coast flying east, the course they held being rather less than a mile off the land. I could only distinguish one small party of Black and a couple of Sandwich Terns, the remainder of the flocks appearing to be entirely composed of Common and Arctic: I remarked that many of the two latter species were in immature plumage." I am unable to call to mind the state of plumage exhibited by the Terns described as immature. The occurrence, however, was noted down at the time the observations were made, precluding all chance of an error having arisen. In June and July I have often shot Common Terns, evidently birds of the previous season, still retaining the dark markings of the first autumn on the shoulder, as well as a few dusky patches on the wings. Arctic Terns also are frequently to be seen in immature plumage resting on the rocks at the Fern Islands, within a few hundred yards of where the adults are engaged in their nesting-operations. Though all particulars have now slipped my memory, I conclude the birds referred to in the note must have been in the stages of those to which attention has just been drawn. I learn from my notes that this flight of Terns continued for some days after the 21st of May. The following entry occurs:—"25th and 26th. Wind south and light, weather dull and fine. Out in the Channel both days. Numbers of Terns still passing; I could only identify one flock of Lesser, the others, as far as could be ascertained, being all Common and Arctic, mostly adult but a few immature. The crew of a fishing-boat informed us that two or three Skuas had been chasing the flocks of Terns; though afloat, however, the whole of both days, I did not catch a glimpse of a single dark-coloured Gull; the pirates were in all probability immature Long-tailed Skuas."

The age at which the Common Tern pairs and nests appears somewhat uncertain; I am of opinion, however, that it is not before the age of two and possibly three years. After the stream of early migrants have passed along our coasts, I repeatedly met with a few of these Terns fishing along the shores of the Channel, exhibiting clear grey backs and apparently adult, with the exception of a dark mark along the shoulder of the wing: the black feathers on the head were also scanty and by no means so well developed as on the adults. As birds in this plumage were to be met with at sea round our shores all summer, though never observed near their nesting-quarters, it was evident that this species does not pair and breed till the perfect adult dress is assumed.

Terns usually move towards the south as autumn draws to a close, and, I believe, are seldom seen after October. While awaiting Ducks at flight-time, however, in the last week of October, 1864, on the shores of the Firth of Forth near North Berwick, a small flock of birds, that I first mistook for Gulls, was observed to flap slowly past the last ray of light remaining in the sky. A moment later, their manner of flight attracting attention, one barrel of a heavy 10-bore was discharged to ascertain the species; that more than one of the unknown had dropped on the sands was clearly audible, and on proceeding towards the spot a couple of birds, which proved on examination to be Common Terns in immature plumage, were picked up, lying dead on the edge of a small pool of water.

In the works of some of the ornithological writers of former days rather curious appellations are bestowed on this species. W. Thompson, in his 'Natural History of Ireland,' published in 1851, refers to this species under the heading of the Common Tern; and adds the names of "Sea Swallows; Pirre (north of Ireland). Skirr at Lambay; Kingfisher at Lough Neagh."

In an edition of the 'History of British Birds,' by the inimitable wood-engraver T. Bewick, published at Newcastle in 1804, I find, under the heading of the Common Tern, the names of the "Great Tern, Kirmew, or Sea-Swallow." William MacGillivray, in his 'History of British Birds,' published in 1852, after using for his description of this species the accustomed name of Common Tern, states that it is also known as the "Pictarne, Tarney, Tarret, Picket, Spurre, Scraye, Kirmew."

The description of the breeding-habits of this species given by MacGillivray agrees with what I have observed myself, and is as follows:—"With us the Terns arrive in straggling flocks in the beginning of May, and soon after betake themselves to their breeding-places, which are sandy tracts, gravelly or pebbly ridges on the shore, rocky ground, or sometimes low rocks. In the latter kind of situation, they make an imperfect nest of bits of grass or fragments of dry sea-weeds; but on sand they merely form a depression." The same author also states, when referring to their eggs:—"The birds usually sit upon them by day, unless in sunny weather, or when they are much disturbed, and always at night, as well as when the air is moist."

T. Bewick, in the old-fashioned type used at the time at which he wrote, gives us the following information concerning the breeding-habits of the Common Tern:—"The female, it is said, forms her nest in the moss or long coarse grass, near the lake, and lays three or four eggs of a dull olive colour, marked with different-sized black spots at the thicker end; it is added, that she covers them only during the night, or in the day when it rains: at all other times she leaves the hatching of them to the sun."

Thompson makes several remarks concerning this species when referring to their proceedings during the spring and rearing their young on the Down coast. He appears to have been an accurate observer; but the assertion that Terns leave their eggs during the day for the sun to hatch them must be an entirely mistaken idea, as I have repeatedly watched, when unobserved, hundreds of these birds sitting on their nests in bright hot weather. The following lines are extracted from his account of this species:—"That the birds do not sit on the eggs during the day, or do so very rarely, is certainly the case at several islands visited by myself. If they did so, they would be hardly less conspicuous than 'snow upon a raven's back;' and hence instinct may prompt them—in localities in which they are liable to be disturbed, both for their own sake and that of their eggs—to absent themselves from their nests in the day-time."

ARCTIC TERN.

STERNA MACRURA.

The difference between the Arctic and Common Tern is at once apparent when a specimen of either species is examined with care; it is, however, by no means easy, unless a close view is obtained, to identify the birds while on wing. A few words drawing attention to their distinctive marks may be of service. The projecting tail-feathers of the Arctic are decidedly longer than those of the Common Tern; the grey tint on the plumage of the breast of the former is also considerably deeper. The mandibles of the adult Arctic Tern are a bright crimson (or perhaps carmine, I hardly know which to term the tint), while dull black points terminate the beak of the Common Tern. In addition, the length of the tarsi, which measure three eighths of an inch more in the Common than in the Arctic Tern, is an unfailing guide.

This species, decidedly more addicted to salt than fresh water, arrives off our coasts either in company with, or at much the same time as, the Common Tern; the habits and manner of obtaining a living followed by the two birds are also exceedingly similar. During my travels in the north I met with but few opportunities for inspecting the breeding-haunts of this species—the large colonies on the Fern Islands, and a few of the smaller nurseries on the shores of the northern Highland firths, where (not unfrequently intermixed with the Common Tern) this species rears its young, being all that have come under my notice. While making observations on the birds at the Fern Islands in 1867, I noticed that Arctic Terns were breeding in immense numbers, three large and two small colonies being established on various parts of this barren and wind-swept group of rocks. While watching the birds at one of their largest nesting-stations, situated near the centre of the islands, my attention was attracted by a flock of over a hundred Terns collected on a low rock, that barely topped the waves, some distance further out to sea; on pulling quietly up, and examining them through the glasses, I ascertained that they were all immature Arctic of the previous year, exhibiting darkly marked backs, and breasts suffused with a deep orange or tawny hue. I ascertained that this conspicuous colouring had led some of the Trinity men belonging to the inner lights, who professed a knowledge of the birds frequenting the islands, to bestow on them the title of Roseate Terns; the egg-collector, however, a nephew of the celebrated Grace Darling, was well aware to what species they belonged. These juveniles, I remarked, made no attempt to approach the islands on which the adult Arctics were breeding, or even to intermix with them when on wing; though occasionally absent on fishing-excursions for the best part of the day, they invariably took up their quarters, when at rest, on one or the other of a few small reefs of rocks near the spot where they first came under my observation.

The old lighthouse that in days gone by warned mariners to avoid the dangers of these rock-bound islands stands near the centre of the group; the light on this antiquated building was in former times supplied by a large coal fire constantly burning on the summit, and the ashes thrown down from above were allowed to accumulate in heaps round the basement, till the surf raised by a heavy gale dashed over the rocks and swept them out to sea. The lower portion of this rough and weather-beaten structure is now used as a storehouse by the custodian of the birds, and at the time of my visit thousands of eggs were packed away in boxes, ready for conveyance to North Sunderland, Holy Island, and various villages along the coast. It is necessary to charge an exceedingly low price for the commonest of the eggs, as the fishermen are by these means induced to become purchasers; otherwise they would land and, searching indiscriminately to help themselves, cause endless destruction. Four eggs of the Cormorant and three of the Guillemot and Lesser Black-backed Gull are sold for a penny, while those of Eiders, Terns, Herring-Gulls, Kittiwakes, and the other species fetch as much as two or three pence each—the former being consumed for food by the native population, and the latter coming into the possession of dealers.

To the best of my knowledge, the following is a correct list of the birds that resorted to these islands for breeding-purposes at the time of my last visit in 1867; with the changes that may have taken place of late years I am utterly unacquainted.

Rock-Pipit.—A few pairs were nesting among the old buildings.

Ringed Plover.—I noticed the eggs of this species on several of these islands; and numbers of the birds were flying about the rocks.

Oyster-Catcher.—Several birds were seen, some in small parties of three or four; Darling, the egg-gatherer, however, was of opinion that only about twelve pairs were nesting this season.

Sheld-Duck.—A few pairs breed on the islands, and several frequent the Magstone, a rock about a mile to the north.

Eider.—Large flocks composed of birds in various stages of plumage, the drakes predominating, were resting quietly on the water in the sheltered bays. The ducks, I ascertained, were sitting on several of the islands; and the men pointed out three or four nests that had been constructed among the rank plants and rough grass against the wall round the old lighthouse, now used for storing and packing the eggs taken for sale. These birds, owing to the constant passing to and fro of the egg-collector and his assistants, had become so confiding that they showed not the slightest fear, even when closely inspected by strangers.

Common Guillemot.—Immense numbers lay their eggs on the summit of the Pinnacles, and a few frequent two or three of the other rocky islets. Several of the ringed form are to be seen intermixed with the crowds assembled on the rocks, when their ranks are closely inspected.

Puffin.—Breeds on some of the islands where the soft mould enables them to scrape out their own domiciles, and also on others where they make use of the rabbit-burrows.

Razorbill.—I only observed a single bird sitting on her egg, and Darling informed me that there was but one pair this season; in former times, however, as at the Bass Rock, they were numerous.

Cormorant.—There seemed to be, so far as I was able to judge, about one hundred nests on one of the rocky islands, the highest being perhaps twenty feet above high-water mark. These birds were stated to change their breeding-quarters every season.

Shag.—This species I learned had always bred at the Pinnacles till the present season; a pair were now about the islands, but it appeared uncertain whether they had as yet commenced nesting-operations.

Sandwich Tern.—There were three large colonies on different islands, their breeding-quarters adjoining those of the Arctic Terns.

Roseate Tern.—Stated by the egg-gatherer to have been not uncommon a few years back. Only a single bird, which I obtained, seen during my visit to the islands; this specimen proved to be a female, evidently sitting at the time.

Arctic Tern.—Very numerous this season; in addition to three large there were two small colonies.

LESSER BLACK-BACKED GULL.—Large numbers breed in colonies on several of the islands.

HERRING-GULL.—Exceedingly scarce; but three or four pairs had taken up their quarters this season.

KITTIWAKE.—A few pairs were nesting in the cracks and crevices of the Pinnacles, immediately below the stations occupied by the Guillemots.

Since my last visit to this part of the coast, I have met with statements, in several ornithological works, to the effect that the Common Tern nested on the islands. As this species, however, escaped identification, though the whole of the Terns were most carefully watched and examined with powerful glasses for three successive days, I do not include it in the list of breeding birds that came under my observation.

The colouring of beaks and legs is seldom alluded to in 'Rough Notes;' the few particulars, however, concerning their various changes to which attention is about to be drawn may not be out of place, as from repeated observations I ascertained that the tints of the soft parts, especially in young birds, commence to fade immediately after death; and it is by no means a certainty to find them accurately described or depicted in even the most trustworthy works on natural history. The plumage, beak, and legs of an Arctic Tern shot on the 9th of August, 1873, in Yarmouth Roads, may be described as follows:—Crown of head black, thickly speckled with white on forehead; breast, back, and wings of the usual adult tints, though somewhat faded and worn; a dark line was also showing across the shoulders of the wings. Beak very deep claret, almost black; legs dark brown with a shade of red. Whether this specimen was an adult undergoing the change into winter plumage, or exhibiting the last stage before arriving at maturity, I am unable to offer an opinion. A young bird shot the same day did not show the tawny orange shade on the breast observed on the juveniles at the Fern Islands in June. The base of both mandibles was a pale flesh-tint, the ridge of the upper, and the points of both upper and lower, black; legs and feet a pale yellowish flesh-colour, nails black. Another in the same stage of plumage, obtained in Shoreham harbour on the 17th of August, 1883, corresponded in every particular in the colouring of the soft parts. Several adults and immature birds of this species procured, for purposes of examination, along the shore near Lancing, on September 1st, 1882, are referred to in my notes; and I find the following description of the colouring of the beak and legs of one of the juveniles:—Ridge of upper mandible and point black; point of lower black, this colour extending halfway up; base of both mandibles pale reddish flesh. Legs and toes pale Indian red; webs a darker tint of same colour; nails black.

LESSER TERN.

STERNA MINUTA.

During the last five-and-twenty years the Lesser Tern has entirely disappeared from several of the breeding-stations to which it formerly resorted in the southern counties along the shores of the Channel, the rage for egg-collecting having doubtless, as in many other cases, been the main cause of this interesting species being driven to haunts on more unfrequented portions of our coast-line. These birds are, however, still sufficiently abundant, breeding at many stations in other parts of the country, to leave little cause for supposing that they stand a chance of becoming a rarity for many years to come. So regardless are these beautiful little Terns concerning their own safety that they hover over those who approach the spots where their eggs or young are concealed, and, in their anxiety to protect them, point out the whereabouts of their treasures. The shingle-banks on either side of Rye harbour were favourite stations of this species at the time I lived in the district, from 1858 to 1862, and numbers put in an appearance early every spring, usually selecting for their breeding-quarters the flat portions of the shingle-banks where the gravel was fine or intermixed with sand and small broken shells. I paid more attention to the colony established on the stretch of shingle to the west of the channel running through the harbour, where from fifty to one hundred pairs were engaged in rearing their young, in the years I visited their haunts. During the breeding-season, immense numbers of Terns of various species, the Lesser perhaps predominating after the spring flights of Sandwich, Arctic, and Common Terns have passed on towards the north, hover round the shallow water in the bay and the pools in the sands, darting down continuously at the small fry that form their food. In still weather, as soon as the tide had risen too high for them to secure their prey, the birds usually settled down and rested on the tops of the poles driven into the sand to hold the "kettle-nets"*, every stake in view at times having its occupant, and each bird sitting with its head turned the same way, facing any light breeze that might ruffle the surface of the water. Occasionally in blusterous weather, with gales of wind and squalls of rain, they would be found in large flocks huddled together on the sands, as usual facing the storms. Occasionally at low water I noticed a few of these birds

* At the time referred to, kettle-nets were set on many parts along the flat, sandy shores of Kent and Sussex, and especially in Rye Bay; not unfrequently immense hauls of mackerel were made, so heavy, indeed, that only half of the take could be saved. Often when brought ashore, and the dealers as well as the whole of the natives satisfied, the remainder had to be carted inland and used as manure or left to rot on the beach. Kettle-nets, to the best of my knowledge, are peculiar to the south coast, none having come under my notice in other parts of the country. A large circular net, enclosing perhaps a quarter of an acre, is set up to the height of about 10 feet, by hop-poles made fast in the sands in a peculiar manner. Spikes cut out of wood are driven through the ends of the poles, and straw-bands then wound around them: a hole is next dug in the sands and the stake with its surroundings placed in the cavity, and the sand returned and pressed down. The circle of net, however, is not quite completed, about 10 feet nearest the shore remaining open; a leader (a net set in the same manner, only in a straight line) is then stretched down from the shore and enters the circle a few feet. The shoals of fish, when making their way along shore, strike the leader and swimming down towards the open sea, enter the circular net and continue heading outwards till left dry by the tide. Frequently there are two and sometimes three "bights," that is, so many sets of circular nets and leaders, all leading straight towards high-water mark. By these means the fish at some distance from the shore are enclosed; it is only where the tide ebbs a long distance on sandy flats that such means for fish-capture can be employed.

flying a short distance up the course of the river towards the old town of Rye, though they generally showed a preference for seeking their prey along the sea-shore.

The majority of the fishermen and gunners along the coast-line of east Sussex give the names of Skerrits or Skiffs to these birds; some of these men, however, know little or nothing concerning the distinction between two or three of the light-coloured smaller species, and speak of them all as Sea-Swallows. I repeatedly remarked that the punt and shore-shooters who gained their living in the harbours and estuaries of the west of Sussex and parts of the adjoining county were well acquainted with most of the Waders, Gulls, and Terns, as well as fowl, being in the habit of procuring specimens for naturalists and collectors, rarities not unfrequently falling into their hands.

During the latter part of August and September Lesser Terns, both adult and immature, commence to make a movement from our shores towards their winter-quarters, and continue, together with those that have winged their way across the North Sea from more northern breeding-stations, to pass along through the Channel towards the west for some weeks. Rough and stormy weather is sure to drive large numbers, while on their way, into the harbours, estuaries, and backwaters along the south coast, and here they remain sheltering from the squalls till the gale has blown over, when their journey is resumed.

The immature birds may easily be recognized during their first autumn by the darker tints of the plumage on the back, which at once attract attention when the juveniles are seen in company with adults.

BLACK TERN.

HYDROCHELIDON NIGRA.

LIKE many another denizen of the meres and swamps of the eastern counties, the Black Tern has been driven from its former haunts, several years having now elapsed since this species reared its young within the limits of the British Islands: at the present time these birds are only seen passing our shores while on their way to and from their breeding-stations; small flocks flying east usually put in an appearance in the Channel off the coasts of Sussex and Kent during the last week in April, and the flight continues throughout May. The earliest date of their arrival on our shores recorded in my journals is April 14th, 1873, when several were seen on Hickling Broad in the east of Norfolk: a light easterly breeze was blowing at the time, and the birds were all flying directly in the face of the wind; on the following day, with a gale and squalls of rain from the same quarter, several small parties of ten or a dozen were still passing.

The minute insects that collect in swarms over the broads and swampy pools in the marshes in the east of Norfolk prove a great attraction to this species on their first arrival in that part of the country. Small parties are to be met with every season, and occasionally I have watched flocks of from fifty to sixty birds engaged in hawking for prey like Swifts; at times they hover over the slades and water-dykes after the manner of a Kestrel, or flap across the flooded portions of the hills with much the same actions as the Marsh-Owl, dipping down now and then for food. On the 28th of April, 1883, with a cold wind blowing from east-south-east, they were especially numerous, and a great difference in the shades of the pale grey colouring of the wings was remarked, some being so light that those who had never met with an opportunity for observing the White-winged Black Tern in life might readily have been mistaken as to the species. Small parties as well as single birds are often seen during the summer months resorting to the Norfolk broads and remaining for several days or even weeks in the district; these stragglers seldom exhibit perfect adult plumage, and are probably birds of the previous year and non-breeders.

The plumage of the young in their first autumn is entirely different to that of the adults; a white brow, cheeks and breast dull grey, back clouded with brown, at once attract attention and proclaim their age. So early as the 26th of July, 1873, I noticed three adults on Hickling Broad undergoing the change into the winter plumage, and two of them were obtained as specimens. All through September I have met with the immature birds passing along our coasts, having observed them on two or three occasions in the Firth of Forth and repeatedly in the channels in the muddy harbours or estuaries of Norfolk, Kent, and Sussex. On their first arrival the juveniles are exceedingly unsuspicious of danger: and while dipping down for small fry in the drains on Breydon Water, near Yarmouth, I often decoyed a specimen or two that was required by means of a three-cornered cork (painted white) flung up into the air. On the lure striking the water the birds instantly sailed to the spot, where they would remain hovering on extended pinions, offering excellent chances for an examination of their state of plumage through the glasses.

WHITE-WINGED BLACK TERN.

HYDROCHELIDON LEUCOPTERA.

Unless this handsome Tern escaped notice in former days, the numbers that pass along our shores have greatly increased of late years. I met with several small parties during the spring in 1871 and 1873; others also that had come under the observation of those well acquainted with all our British species were reported on various parts of the coast.

A few abridged extracts from my notes for 1871, 1872, and 1873 will impart all the knowledge concerning the habits of these birds that I am enabled to give. In 1871 these Terns were seen on Breydon mudflats, and a couple of years later on Hickling Broad.

"May 26, 1871. After a heavy thunderstorm over the town (Yarmouth) during the night, the morning broke exceedingly dull and overcast, rain falling heavily. There was just sufficient light on reaching the 'lumps' in the gunning-punt to make out five Terns pitching in the channel above us; though the birds did not approach within range, a closer view which was obtained of one of their number led to the belief that the whole were White-winged Black Terns. As the daylight increased they worked further up the flats, and finally four settled between two brightly plumaged Grey Plovers sitting about six feet apart at the side of 'Bessie's drain.' From the sudden manner in which the birds wheeled round and alighted, it is probable they were attracted by the conspicuous colouring of the Plovers, their black breasts with the edging of white corresponding almost precisely with their own appearance. Though one remained hovering overhead, the four happened to have settled so conveniently at the moment we came within range, setting slowly up the drain with the flood-tide, that the chance was not to be lost, and the whole party were secured by a charge of small shot from the big gun. The remaining bird circled round for a time at a great height, evidently reluctant to leave its companions, but eventually darted off towards the north-east, where a flock of Waders, disturbed by the shot, were wheeling over the flats. The specimens secured proved to be two males and two females, in the finest summer plumage. Possibly the storms of the previous night may have carried these birds out of their usual course, though it is highly probable that stragglers visit our shores more frequently than is supposed. One was seen two days previously flying, in company with a small party of Common Terns, across the flats; though the decoys (which seldom prove of use in spring) were flung up, they continued on their course, passing away towards the north-east. When first observed, these Terns were plunging down headlong in the channel, apparently in pursuit of prey, though owing to the imperfect light it was almost impossible for them to have discerned any small fish. Black Terns in spring frequently, while on wing, take an insect from the water, but I do not remember to have watched them darting down and seeking food below the surface; this action repeatedly performed first attracted notice and led to their identification."

In the spring of 1872 I was in the south of England, and though almost daily at sea did not meet with the species; a pair were, however, seen on Breydon by one of the gunners who had been present when the specimens were procured the previous year.

"May 28, 1873. Wind north and cold. Having heard that several 'Dars'*, whose description answered exactly for this species, had been noticed for the last four or five days frequenting Hickling Broad, I rowed round the water in search of the strangers. The keeper who gave me the information stated that in his fifty years' experience of the Broad he had seen nothing like them before; as the man was a good authority on all the native fowl, waders, and sea-birds, it is probable that their visits to this locality were unfrequent. I had not pulled above a quarter of a mile when a fine old male was detected hovering round the edges of the hills, and a few hours later a party of six or seven were met with flying over the Broad. Thousands of Sand-Martins skimmed hither and thither across the surface of the water, darting at the insects, and the Terns, immediately joining a large swarm, continued in their company for some time. In no single instance did I see them dash into the water as observed on Breydon, and in all probability food is only procured in this manner on salt water; during the whole of the afternoon they remained hawking for insects in precisely the same manner as the Martins. After watching their actions for some hours, several of their number often approaching and sweeping round the punt within the distance of three or four yards, I procured without difficulty as many specimens as were needed. On the following day several small parties came in view beating to windward over the Broad; a fresh breeze from the north-east was blowing, and after a few turns they worked away in the face of the gale; the wind being too strong for the insects to show themselves, the Martins were absent during the whole of the day. Three or four of the birds seen towards evening were more strongly marked on the breast with white than those previously observed; these doubtless exhibited some of the more immature stages of plumage."

* Terns are generally known among the gunners and marshmen in the east of Norfolk as "Dars" or "Daws," some being termed blue and others black, according to the tints of their colouring.

LITTLE GULL.

LARUS MINUTUS.

It occasionally happens that Little Gulls in considerable numbers make their appearance off various parts round the coast-line of the British Islands: though but few birds came under my observation, I have seen fresh-killed examples of this elegant species of all ages and in every stage of plumage.

While on Horsey Mere * in the east of Norfolk on the 21st November, 1871, I noticed an immature bird skimming over the water, and remarked that its actions (with the exception that the headlong plunge was omitted) appeared much to resemble those of a Tern. Flapping slowly head to wind, and dipping down occasionally, though without touching the surface of the water, this small Gull made its way along one side of the mere; having reached the entrance of the river, it turned, and dropping back more rapidly towards the east end, again worked slowly over the same course. The bird appeared to be searching for food, but was unable to find any; possibly a deep freshwater broad was hardly a suitable feeding-ground for this species. After watching its movements for some time, but little difficulty was experienced in securing it as a specimen, the confiding little stranger having repeatedly passed the boat within the distance of twenty yards.

The bird proved to be in the usual immature plumage, the dark bars on the wings (somewhat similar to those on the juvenile Kittiwake) being exceedingly conspicuous while flying. I noticed that the rosy hue on the breast was far deeper than that usually depicted in the coloured plates of even the adults; the tints, however, commenced to fade shortly after the death of the bird, and by the time it was preserved had entirely vanished.

* A large sheet of fresh water in a marshy district about a mile from the sea-coast.

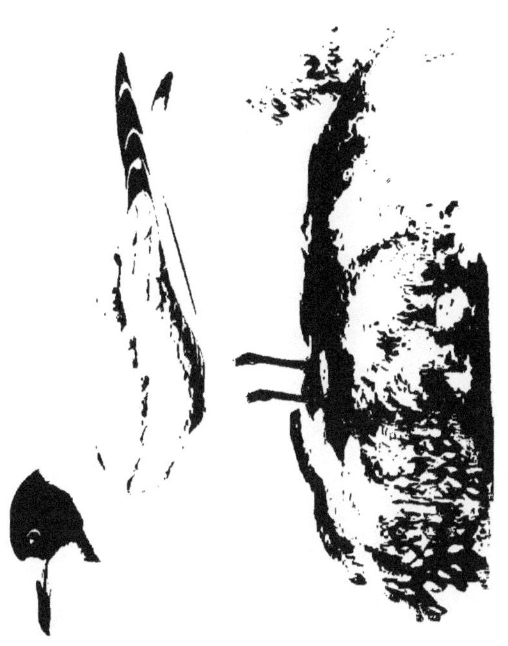

BLACK-HEADED GULL.

LARUS RIDIBUNDUS.

I HAVE met with the Black-headed Gull all along the eastern coast-line from Caithness to Sussex, as well as off most counties bordering the sea-shore. This species does not appear to have perceptibly diminished in numbers, though many of their haunts of late years have been encroached on by innovations, such as drainage or railroads. Usually nesting on preserved grounds, this Gull does not suffer from the persecution which birds breeding along the sea-coast are exposed to, and when driven from their nesting-places by the reclaiming of swamps and marshes for farming-purposes they usually find little difficulty in selecting suitable quarters at no great distance.

This species is without doubt somewhat nocturnal in its habits; while staying at the small village at Canty Bay, opposite the Bass Rock in the Firth of Forth, in the autumn of 1871, I often noticed these birds collecting towards evening along the sands just above high-water mark, and on making further investigations ascertained that they were engaged in capturing the sand-hoppers, minute marine insects somewhat resembling small shrimps, that swarm in thousands under the dead seaweed thrown up by the tide and lying in heaps along the shore. The humming noise that these tiny creatures emit when disturbed is far louder than would be supposed, and resounds on all sides if the shelter under which they are concealed is in any manner interfered with. While in quest of specimens of the Black-headed Gull in its various stages of plumage, I took the marks one evening where a party of these birds generally alighted to feed, and approaching after dark when they were busily occupied, succeeded in knocking over half a dozen with the two barrels of a 12-bore breech-loader. The slain, when picked up, proved to be mostly juveniles, though there happened to be one or two almost adults; all, however, were acceptable as specimens exhibiting the different stages through which they pass. These Gulls may also be seen towards dusk about the Broads in the east of Norfolk, flying backwards and forwards across the marshes and darting down at the insects hovering round the brambles on the banks, and the reeds and rushes along the dykes. The light colouring of the ghost moth is exceedingly conspicuous at such times, and I frequently remarked that they were greedily snapped up by the Gull; probably other insects are also captured in localities where these moths are not to be obtained.

Very large colonies of these birds are occasionally found breeding in company; at Loch Doula near Lairg in Sutherland, three islands in the loch were almost covered with nests at the time of my first visit in June 1868. There were also some hundreds of nests on a waving bog to the south of the loch, to which it was almost impossible to make one's way, so scanty was the covering of soil that had formed above the decomposing mass of rotten reeds and mud on which the Gulls had placed their cradles. A few Common Gulls were also breeding near at hand, their nests being situated here and there in the long heather round the edge of the loch, but not intermixed with those of this species.

On my first inspection of this exceedingly interesting loch, I discovered there was no boat on the water; this deficiency, however, was easily remedied, as on returning a few days later one of my portable india-rubber boats that have done so much in remote districts was brought to the spot. On the 15th of June, after exploring all the islands and the nearest portions of the swamps to which I could make my way, as well as obtaining a few young of this species in the down for specimens, I turned to paddle to the shore where the keepers were awaiting my return. Nearly the whole of the juveniles that were still unfledged had now joined into one large body of perhaps twelve or fifteen hundred, and were swimming in front of my boat, when I noticed four small sharp-winged birds fly up from the surface of the water in front of the swarm, and after skimming round two or three times, alight on the shore close to where the men were making preparations for my arrival and the repacking of my craft. I had but little doubt the tiny strangers were Red-necked Phalaropes, and one of the keepers who came from my shootings at Innerwick in Glenlyon, in the west of Perthshire, called out that such was the fact before I landed. This man had been present one evening the year before in September, when I obtained a Grey Phalarope in a marshy field near the banks of the river Lyon, and was well aware, from having looked over a work on natural history, that there was another species. The birds took little notice of us after I came on shore, and showed not the slightest signs of alarm, running up occasionally while feeding within half a dozen yards, and picking about among the stones by the water-side. At this date I imagined the two pairs must have selected their breeding-quarters, and consequently no attempt was made to interfere with them before their nest had been detected. After remaining till our portable boat had been emptied of the air and stowed away in its case, and our refreshments were consumed, we moved to a low ridge at the distance of about a couple of hundred yards, so that a watch might be kept on the Phalaropes through the glasses, to ascertain if any of their number made off towards likely spots where their eggs might be concealed. As the birds all kept together and gave no signs of leaving the loch-side, I came to the conclusion that they could not yet have commenced nesting-operations, and determined to take advantage of the first opportunity to obtain a shot. On reaching the shore one pair was observed flying round at a low elevation over the water, and just before I came within range the others rose and joined them; the whole party then swept upwards to a higher elevation than I had previously noticed and made straight off towards the north-east. Though a constant watch was kept for a fortnight or three weeks with but few hours' intermission, both night and day, these wanderers were never observed again in the same locality, and there is little doubt they were only on their way to more northern breeding-quarters. I have always regretted that these Red-necked Phalaropes were allowed to escape, and should have secured them at once, only the fact that a nest and pair of birds had been previously obtained at a loch but a few miles distant induced me to believe that these pairs might already have taken up their summer-quarters.

Early one morning a week or two later, just as the mist was clearing off at daybreak, I reached the shores of the loch, and taking up a position among some large dead stumps of trees proceeded to wait and make observations in hopes that our old friends the Phalaropes might again put in an appearance. I had halted on the south side facing the bog on which the Gulls had nested on the mainland, and was watching one or two that appeared excited and were swooping round uttering loud cries, when a wild cat came in view picking its way stealthily over the swamp with a young Gull in its mouth. I was enabled to see through the glasses that the bird, which was partially feathered, was dead, and it was also obvious that the feline marauder had not discovered my presence, a few patches of dead reeds affording ample concealment. Unwilling to disturb the birds on the loch by a shot, I allowed the animal to depart unmolested with its prey, and it was soon lost sight of, having turned towards the north, its quarters being probably among the hills near Ben Armine.

At a rush-grown loch connected by a small stream, navigable for my punt, with Loch Slyn near Tain in the east of Ross-shire, there was another breeding-place of this species, where immense numbers of nests were scattered about among the beds of rushes where the birds could find situations sufficiently dry and raised above high-water mark when floods occurred. On the 13th of May, 1869, I took an egg of a pale blue shade, resembling the tint of the egg of a Wheatear, without any markings; the other egg in the nest proved to be of the ordinary colouring. Repeatedly at this station and also at Loch Doula, as well as in other Highland colonies, I noticed eggs in the nests of very light tints of greenish brown and sometimes blue, but never met with others without some few spots or clouded markings.

While collecting in the Western Highlands in the summer of 1868, I thoroughly explored the wild district around Loch Maree, making my way into many of the remotest glens: during my wanderings I was informed by the keepers at Letterewe that these Gulls had regularly nested at Loch Garvaig *; this lonely sheet of water was situated near the foot of the western slopes of Ben Slioch and shut in on all sides by high mountains; it was discovered, however, on visiting their usual breeding-haunts on the 5th of June, that there was not a Black-headed Gull to be seen about the loch. The cause of their disappearance was quite unintelligible, as they had never suffered the least persecution, nor had their nests been robbed to any extent: it was doubtful, indeed, if even a few eggs had been taken, so inaccessible was the spot the birds had selected, and such an unfailing supply of the eggs of the larger Gulls being always at hand on the islands of Loch Maree to supply the wants of the crofters. Shortly after our departure on this occasion, a heavy storm accompanied by a deluge of rain broke over the district, and flooded many of the breeding-quarters of Gulls, Ducks, and Divers, destroying thousands of nests. If one species could possibly have been impressed by an inkling of impending danger, and induced to change their quarters, why should not the others have received a similar warning? The previous year I became aware of the disappearance of the Sand-Martins from an island in the river Lyon, in the west of Perthshire, and early in June a terrible thunderstorm with a heavy downfall of rain among the mountains in the upper part of the glen caused a spate that flooded the whole of the island on which the birds usually remained to rear their young, to the depth of between two and three feet†.

In the east of Norfolk I learned from the marshmen who had passed their lives fishing and shooting on the Broads about Potter Heigham and Hickling that Black-headed Gulls had formerly bred on one or two of the hills surrounding the latter piece of water, though they had now (1870) entirely ceased to remain and construct their nests. Numbers continued to fly about the Broads during the summer months; these, however, all exhibiting full adult plumage, were without doubt breeding birds from Scoulton Mere or some other smaller stations induced to wing their way to these attractive waters by the abundance of food to be obtained. After the bird-protection act came in force, and the shooting in spring of certain species was prohibited by law, many more of these Gulls put in an appearance, and I heard that two or three pairs had returned and taken up their quarters on one of the hills adjoining Hickling Broad. The men who made these statements were uncertain whether the young had been reared or even hatched out, and I could gain no further information on the subject beyond the fact that no nests were built the following year. In 1883 I remained for some months in this part of Norfolk, and in the end of June noticed a nest on "Swimcoats," a hill on the west side of the Broad; this was built in one of the slades among the strong stems of a large tuft of rushes, at a height of about fourteen or sixteen inches from the ground, and had probably been constructed at the time when heavy rains having fallen, a flood had raised the water above its usual height on that part of the marsh. There were neither eggs nor young, and it would have been strange if the Black Crows that frequented the plantations round the Broad and daily hunted over the hills had allowed the contents of a solitary nest to escape their attentions. Again, in 1885, I closely searched the old haunts of this species, but discovered no signs of their presence; the natives also had failed to observe more than the usual complement of visitors from other quarters hovering over the waters of the Broad or flitting round the pools and slades on the hills.

Never having kept Black-headed Gulls in confinement for any length of time, I cannot state with certainty the age at which the perfect adult plumage is assumed. The various stages in which this species is to be seen

* Garvaig is the name given to this loch in Black's map of Scotland; by the natives of the district, however, I remarked that it was always spoken of as Loch Gararach.

† This circumstance is referred to in 'Rough Notes' under the heading of the "Sand-Martin," on pages 1 and 2.

and the numbers to be observed in company along the coast-line in summer render it probable, however, that they do not reach maturity till three or possibly four years old.

There is little doubt that the two birds shot at Canty Bay in East Lothian in August 1874, and depicted on Plate I., represent the plumage of the first and second autumn. The colours of the soft parts of these two specimens were as follows:—"The upper and lower mandibles of the juvenile were flesh-tinted with a black point on the upper; legs and feet a pale livid flesh-colour. Both mandibles of the older bird were red, with a black point on the upper, and a light red circle round the iris. The legs and feet were also of the same tint of red as the mandibles.

The adult figured on Plate II. flew rapidly past my gunning-punt on the 18th of January 1881, about half a mile at sea in the Channel off Lancing, in West Sussex, and his striking appearance having attracted my attention, he was knocked down by the shoulder-gun. This strangely marked bird exhibits the singular manner in which the change from white to black on the head is occasionally effected. The upper and lower mandibles were a dark purple tint, with a tinge of crimson-lake near the base; the circle round the eye was very conspicuous and of a bright mulberry hue. The legs and feet showed a colouring of dark purple.

COMMON GULL.

LARUS CANUS.

Though seldom, if ever, seen gathered into such immense flocks as the Lesser Black-backed Gull or the Kittiwake, the Common Gull is to be met with at one season or another all round our coast. In most of the southern and eastern counties with which I am acquainted this species is more numerous in winter, the stragglers (for the most part showing various immature stages) that have remained during summer being joined as autumn approaches by the adults and young birds of the year. For some months their favourite quarters are in the vicinity of the outlets of the sewers of the fashionable watering-places or about the harbour-mouths of seaport towns. Here, with swarms of other ravenous seafowl and at times a noisy party of Grey Crows, the filth and garbage that floats in the tideway is eagerly sought after, and many are the contentions that arise over the more coveted portions. Gathered into large bodies on the adjacent sands, or floating quietly at sea during the flood, they patiently await the turn of the tide: shortly before high water they commence to show signs of life—first one and then another rises on wing, and after hovering for a time in the direction from which the anticipated feast will flow, again join the main body *. In severe frosts or when protracted storms have worn out the birds and rendered them careless of danger, I have frequently watched numbers flitting round the fishing-boats and vessels moored alongside the quays in the harbours of the east coast ports. During the terrible easterly gale that broke over the coast of Norfolk early in November 1872 the Gulls frequenting the shore near Yarmouth suffered greatly from the severity of the buffeting they had undergone, large numbers being carried in a helpless condition up the river, while others where driven many miles into the country.

Common Gulls are frequently to be met with at long distances from the coast; during autumn and winter they may be seen following the plough, often in company with Rooks, or picking over any fresh-turned soil. On the 3rd of April, 1883, I remarked large numbers (all, with but few exceptions, exhibiting the adult plumage) scattered over the cultivated land on the downs near the Dyke hill in Sussex; they were busily searching for food, in many instances within a few yards of the road, paying little or no regard to the traffic. Unless driven in by gales these birds are by no means such regular visitors as might be expected to the Norfolk broads, the Gulls usually seen in those flat and marshy districts being the Lesser Black-backed, the Herring-Gull, and wandering parties of the Black-headed.

During summer vast numbers of this species may be noticed in various parts of the Highlands frequenting the shallows of the rivers, occasionally wading knee-deep and searching among the stones or dipping down here and there while on wing. Though it is probable that other prey may fall to their share while seeking for food in such spots, I am convinced that the silvery little smolts (the young of the salmon) form almost their sole diet. In the more northern counties I repeatedly remarked that in several instances Arctic Skuas resorted

* I noticed during several winters that the whole of the commoner species of Gulls were represented in the flocks frequenting the Roads off Yarmouth and Lowestoft, the Lesser Black-backed and the Common Gull being usually the most numerous.

to the same localities, the hapless Gulls being perpetually plundered and harassed by those dashing robbers*. Considering the numbers of Gulls passing the summer on the moors, the quantities of young salmon destroyed must be enormous. Young birds are also occasionally taken, though but few instances of such depravity have come under my notice. While crossing the moors near Loch More, in Caithness, on May 31, 1869, two adult Common Gulls were seen quarrelling over some small and helpless object; on approaching the spot the birds rose from a downy nestling of the Golden Plover, and within a few yards I detected a second, partially devoured; both of these innocent victims were still warm, plainly indicating that the murders had only recently been committed.

The breeding-haunts of this species are in many instances situated at long distances from the sea-coast. To the north-west of Ben Slioch, the highest mountain in Ross-shire, a colony of some hundreds is established on the islands as well as on the rough ground surrounding a hill-loch situated in the midst of some of the wildest scenery in the Western Highlands. It is seldom this desolate spot is visited save by the deer-stalker or a wandering shepherd; and the utter disregard of their natural enemies shown by even some of the most wary species nesting in the vicinity was exceedingly striking. After a short clamour had been raised on our first approach the Gulls settled quietly down and but slight heed was taken of our movements for the remainder of the day. Black-headed Gulls, I learned from our guide (a keeper from Letterewe), had formerly bred here in great numbers, though not a bird was seen during the hours we spent by the loch-side; and it was evident the colony must have shifted their quarters. A pair of Black-throated Divers were nesting on the island, and while visiting their haunts in the india-rubber boat (the only kind of craft that could be transported to such inaccessible spots) I noticed both birds on the water. Shortly after returning to the shore, the female was again seen on her eggs, where she remained sitting till we left the loch, the male repeatedly showing himself on the water during the afternoon within half a gunshot. The majority of the nests of the Gulls were placed near the loch-side or on the adjacent moorland; a few, however, I remarked were scattered among some rough tussocks of grass at a slight elevation above the level of the ground. It is probable that many of the former must have been carried away by the rush of water from the swollen burns during the storm that occurred the following night—the 5th of June, 1868. Shortly before dark the weather assumed an exceedingly threatening aspect and at length a heavy downpour of rain set in, lasting without intermission for several hours; the smallest streamlets increased to mountain-torrents, and the lochs rose rapidly to such a height that thousands of the eggs of Gulls, Ducks, and Divers must have been destroyed. Owing to the length of the journey and the roughness of the track, I did not return to the loch to ascertain if the nests of the Gulls that had attracted my attention escaped the effects of the flood; judging from observations made in the glens on the south side of Loch Maree, I believe they were sufficiently raised to save their contents from contact with the water, though the whole of the nests by the loch-side must inevitably have been submerged.

Far up towards the west, in the remotest part of Glenlyon, in Perthshire, a large colony of Common Gulls has been established for many years, though latterly the birds have decreased in numbers. The lochs of Roro, where this Gullery is situated, are far nearer to the western sea-coast; the birds, however, invariably make their way in spring up the Tay and following on through Glenlyon reach their mountain-haunts by crossing the moors near the course of the burns falling from the lochs. It is usually the beginning of April before any numbers are seen about the Lyon; if looked for they will then be found by the river-side searching diligently for smolts or small salmon while gradually working on towards their summer-quarters. Occasionally I have watched small parties settled quietly resting for hours near the beds on which the salmon had previously

* I regret that no information was received from the taxidermist concerning the contents of the stomachs of any of the Common Gulls procured as specimens at their summer-quarters. None of the Arctic Skuas, however, were found to contain less than two or three smolts, still in an undigested state. Since these Skuas seldom, if ever, capture fish for themselves, depending almost entirely on the exertions of the unfortunate Gulls for their supply of food, there can be little doubt as to the predatory habits of the latter.

spawned, though for the most part they were to be observed flitting hither and thither where the current swept rapidly over the shallows. On the westernmost of the two lochs is a small island (said by local tradition to have been the residence of a far-famed freebooter), and here were some two hundred nests at the time of my first visit in 1866. Dead strands of grass with a few fine fibres of root had been employed by the birds in the construction of their cradles, which were inserted in every available hollow on the rough ground, as well as in the cracks and crevices in the slabs of the moss-grown rocks. Here and there nests were to be seen built among the roots and fallen trunks of rotten timber, partially concealed by the fronds of the large and spreading ferns that flourished luxuriantly in the moist and peaty soil. In a few instances a situation had been chosen in the old and weather-beaten birches where the limbs branched out from the main stem; none, however, were placed at a greater altitude than about four feet from the ground. Though the lower portion of the trees still exhibited foliage of the brightest green, the topmost twigs and branches had been killed through the Gulls constantly perching. The snow-white birds resting motionless on the dead and withered limbs, or steadying themselves with flapping wings while taking up a position *, gave a somewhat ludicrous appearance to the busy scene. At the same time hundreds might be observed sweeping round the clump of trees or dotted here and there over the rocks and stones, while, if carefully looked for, others, half hidden by the luxuriant vegetation, could be detected sitting quietly on their nests.

During the years I spent in the west of Perthshire this breeding-station was carefully protected by the proprietor, every care being taken to prevent intrusion on the haunts of the birds. After a few eggs had been procured at the commencement of the season, the only boat on the loch was removed, and communication with the island cut off till after the young were on wing. Shortly after the boat had been carted away in 1865 a pair of Grey Crows built their nest in the branches of one of the stunted birch trees and fared sumptuously on Gulls' eggs till the craft was brought back and an effectual stop put to their depredations.

In all of the northern counties of the Highlands with which I am acquainted there are several breeding-stations of these Gulls, for the most part situated out on the barren and open moors. About Loch Doula, near Lairg, in Sutherland, numbers of this species take up their summer-quarters; the whole of the available space beneath the shelter of the stunted bushes on the wooded islands being, however, appropriated by the swarms of Black-headed Gulls that resort to this locality, the Common Gulls are banished to the rough ground round the edges of the water and a swamp on the south side; here I have never witnessed them attempting to settle on the trees, and it is but seldom that the Black-headed will make more than a clumsy effort to balance themselves for a few moments on the waving branches.

On the western coast of the Highlands I repeatedly observed nests of this species placed among the ledges of rock a short distance above high-water mark; in some instances a single pair will be met with, at others a small colony rear their families in company; these birds, however, do not crowd their offspring together after the fashion of Kittiwakes. Common Gulls are stated in various works to nest in many parts of the country in the high cliffs overhanging the sea-coast; unfortunately I do not happen to have explored any such spots to which they resorted, and consequently am unable to give even the slightest information on my own authority concerning their breeding-habits in localities of this description. With regard to the nesting-station of the Common Gull reported to be situated on the Bass Rock, I have no hesitation in asserting my belief that some error must have occurred. I am well acquainted with all parts of the rock, and having closely examined the grassy slopes near the summit, as well as the ledges in the face of the cliffs, repeatedly during several seasons, it is extremely unlikely that the species could have escaped notice. The fishermen of Canty Bay, who have had charge of the fowling-operations on the Bass for many years, possess no little knowledge concerning the sea-birds and their habits, and from them I could never learn that this Gull had bred upon the rock.

Though Common Gulls are by no means so ready as many other species to attack any feathered stranger

* Occasionally the Gulls experienced considerable difficulty in gaining a footing, but when once settled were perfectly at ease.

that ventures near their breeding-quarters, I witnessed an amusing scene at Loch Inver, in Sutherland, one evening in June 1877. Several pairs of Gulls frequent the lonely rock-bound coast to the south-west of the Loch during summer; one nest, however, was placed just above the wash of the tide, but a short distance from the village. While watching the two old birds fishing along a sandy bay in the immediate vicinity of the low-lying ridge on which their young were located I noticed a Long-eared Owl flap slowly towards the water. Evidently disturbed from its shelter in the pine-woods before the accustomed hour, dazzled by the light and apparently at a loss which way to turn, its uncertain and wavering flight speedily attracted attention. Instantly the male Gull with loud screams dashed after the intruder and buffeting the bewildered bird repeatedly forced it out to sea; roused by the outcry, a fresh contingent of Gulls shortly arrived and at once joined in the attack with the greatest fury. The Owl, after having been driven over the centre of the loch, at length rose high in the air, followed by a couple of the most inveterate of its pursuers, and not till it had disappeared from view among the hills to the north did the Gulls return to their quarters and was peace reestablished. On no other occasion have I seen an Owl rise to such an altitude; at one time it wheeled round in large circles at the height of at least three hundred feet.

The nestling, immature, a few of the intermediate, and the adult plumages of this species have been described; the writers, however, have failed to state how many years elapse before the perfect mature dress is assumed. I never kept the Common Gull in captivity for any length of time, and consequently have been unable to follow all its changes; but the bird in all probability is not under three or four years of age before it pairs and nests. Under date of August 1st, 1873, my notes contain the description of a Common Gull shot on Breydon mudflats which I should judge to be just over two years old:—"Head and neck white, slightly speckled with grey; back clear blue; wings blue, intermixed with brown; tail white with a black bar; iris brown; beak a pale blue-green, brown at the point; legs and feet pale blue-green."

Under various titles this species is mentioned by different writers; one name made use of, "the Green-billed Gull," is perfectly correct as to the bird in the adult state, though the author omits to draw attention to the fact that, while immature, the bill is flesh-tinted with a dark horn-colour towards the point, which, again, is lighter. Bewick, when giving a description of the Winter Gull (*Larus hybernus*, Linn.), evidently neither more nor less than one of the immature stages of *Larus canus*, bestows on the species the quaint appellation of "Winter Mew" or "Coddy Moddy."

LESSER BLACK-BACKED GULL.

LARUS FUSCUS.

As far as I have been able to judge from observations made in all parts of our islands, and also on the surrounding seas, this species is the most numerous of the British Gulls, breeding on various rocky portions of the coast, and in colonies of larger or smaller size on the islands in the inland lochs as well as on the open moors in the Highlands. To describe the whole of the nesting-stations that have come under my notice is needless; but a few may be mentioned to give an idea of the nature of the country in which these handsome Gulls pass the summer.

The islands on Loch Maree, in the west of Ross-shire, are resorted to by thousands of pairs: here they are permitted to rear their young in comparative peace, as boats are (or rather were, for I have not visited the spot for some years) scarce on the loch, and it is but seldom that the country people are able to reach the islands to obtain their eggs. Those who have only viewed this beautiful loch under the influence of a bright sky and a gentle breeze, would never credit the fury of the squalls that at times gather among the surrounding hills and burst with but scanty warning over its surface. On one occasion, when I had sent the previous day to the keeper to ask for the use of the boat, it was discovered, on arriving at the spot, that three girls had come down from the hills in hopes of getting out, to procure a few baskets full of eggs. After landing them on the islands where the Gulls were most plentiful, we proceeded to search for the nests of Geese and Divers, or other rarities that might fall in our way. Though the early morning had been fine and still, by noon it had clouded over, and rain and wind set in. While paddling among the islands in the india-rubber boat on the watch for Geese, and inspecting the deserted haunts of the Osprey, I narrowly escaped being caught in the first outburst of the storm *. For some hours we delayed our return voyage in expectation that the weather would moderate; at last, while attempting to reach the shore with a large cargo of eggs gathered by the lassies, we were struck by a squall that came roaring across the loch with a blinding cloud of spray, and driven back on one of the islands, the breaking of an oar sending two of the crew to the bottom of the boat, where they rolled about with the eggs, now being dashed from side to side. On working our craft into a sheltered bay and landing our terror-stricken passengers, we were forced to work hard to repair the damages received, and by the time our defects were made good, the storm had abated sufficiently to make a second attempt. The girls, who had crouched at the bottom of the boat, presented a most ludicrous appearance, being drenched to the skin in a mixture that resembled egg-flip, the whole of their spoil, consisting of several hundred Gulls' eggs, having been smashed and beaten up into a kind of custard with the water that had broken on board. Some meat and drink, and the attentions of two or three sturdy keepers, eventually put fresh life into the disconsolate maidens; but when they took their leave in the gloaming it was hard to recognize in the three limp and draggled forms the bright-looking lassies that had met us in the morning.

* This incident is referred to under the heading of the Osprey, on pages 2 and 3.

At the time of my last visit to Caithness large numbers of this species nested out on the dampest of the flats in the central portions of the county; many of these colonies must, I should imagine, have been broken up by the construction of the railroad. Much of the wild moorland to the north of Loch Shin in Sutherland has been brought under cultivation by the help of steam-ploughs, and doubtless here also the Gulls that bred in the localities have been banished from their summer-quarters. On one occasion, in June 1868, while spending the night on the hills near Loch Beannoch, a short distance to the north of Loch Shin, I remarked that these birds were continually passing over the moors towards the west, where a few colonies existed in those days. There was not half an hour during the night that two or three were not seen, all holding the same course; doubtless these birds were returning from a lengthened flight to the salt-water firths or the open sea. The rocky ledges along the barren northern and western coast-line, and the dreary stretches of marshland surrounding the inland lochs and pools in the outlying islands, to which the Lesser Black-backed Gull resorts, are mostly free from intrusion, and unless a few eggs are collected by the families of the crofters for food, the birds suffer little annoyance during the breeding-season. At the Bass, the reasons assigned for the diminution of the Herring-Gull have also affected this species; on the Fern Islands the protection afforded by the presence of the egg-gatherers kept up the stock to the usual standard during the years I received information on the subject.

I am aware of no breeding-stations about the upper waters of the Lyon, in the north-west of Perthshire; several of these Gulls were, however, taken regularly every spring in the vermin-traps set on the moors during the years the Innerwick shootings were in my hands. These birds appeared to follow the course of the river up from the Tay, passing usually at a higher elevation on the hills than the Common Gulls, not one of which was ever captured by these means. The carnivorous propensities of this species may be judged from the fact that while killing down the vermin in 1866, several were taken at baits consisting of fox-cubs, cats, and blue hares; eggs also proved attractive on two or three occasions the following season. In the more northern counties of Ross, Sutherland, and Caithness their depredations cause considerable losses to the game-preservers, eggs and young of Grouse, Plovers, and Wildfowl suffering severely from their attacks; they appear, however, by no means so destructive to the smolts as the Common Gull.

At the time of the bringing out of the Sea-Bird Act it was stated that Gulls were of great assistance to the fishermen, by indicating the whereabouts of the fish, and so guiding them to the best spots for shooting their nets. This all looks very pretty and interesting in print, but I have yet to learn that the Gull is a favourite with the seafaring population. When the shoals of mackerel arrive off the south coast in the spring, hundreds of seine-boats are engaged in watching for the fish to come to the surface; as soon as they are sighted the crews row rapidly to the spot, and shooting the net round them, frequently enclose large numbers. Should any Gulls, however, be near at hand, their sharp eyes are sure to detect the first ripple on the water, and dashing down into the middle of the shoal, the fish are driven to the bottom, and the men who may have rowed hard for half a mile or more, and possibly paid out a portion of their net, find their time and labour thrown away, while the mischievous birds, with a derisive scream, sail off to repeat the performance at the earliest opportunity. While watching these proceedings off Brighton and Shoreham, I have often been requested by the crews of the boats to shoot the Gulls, the men declaring that what with the Bird-Act and the Gun-License they were unable to help themselves, being forced to stand quietly by while the birds snatched the bread from their mouths. The number of these Gulls that congregate in the North Sea during the herring-season is enormous; here, again, they cause great loss to the fishermen. I have been assured by the masters of some of the luggers that they have not unfrequently been deprived of a last of herrings, and occasionally up to four or five times that amount, by their depredations. As a last is ten thousand fish *, the quantity might seem

* A "last," though spoken of as ten thousand, in reality contains 13,200 herrings. When the boats bring their cargoes to the wharf at Yarmouth, the fish are counted out six score and twelve to the hundred; mackerel in this port are reckoned at six score to the hundred, and

incredible to those who have never had an opportunity of watching a large flock of these birds gathered round a boat making a good haul. The numbers these robbers consume is small compared with those they bite and shake from the nets. I have repeatedly observed at least two or three thousand of the larger Gulls (the present species apparently numbering about ten to one) attacking the nets of a single boat when a heavy catch has been secured. While a pause occurs in the operations they swim in compact bodies around the buoys, eagerly watching for a chance; when the capstan is again in motion and the nets, glistening with countless herrings like a stream of silver, come gradually in view, the whole mass at times rise on wing, and seizing hold of the lines in their beaks attempt to shake out the fish. The small boat is occasionally sent to drive off the birds, but if disturbed from one part of the nets they rapidly commence operations on another. After the boats have finished hauling, the birds are usually satisfied, resting in large bodies upon the water for the remainder of the day; it is by no means uncommon to meet with a flock that extends a mile or two in length. By 9 A.M. the majority of the boats have their herrings on board: should the nets, however, become entangled, many hours are often spent in the endeavour to set them free; such mishaps never escape the notice of the Gulls, thousands continually circling round or dashing down to the water till the difficulty is at an end. So long as there is a chance of food they will remain in attendance. One afternoon in October 1872, we fell in with a couple of luggers from Yarmouth and Lowestoft, about twenty miles outside the Cross Sands, whose nets had been dragged into an inextricable confusion by a steamboat. The immense flock of Gulls, appearing at a distance like a swarm of bees, first attracted our attention; and the following day at noon the same two boats were again passed still engaged in recovering their damaged nets, with much the same number of birds surrounding them.

For the greater part of the information given in these pages concerning the Gulls and Skuas in the North Sea, that is not derived from personal observations, I am indebted to the brothers Thomas, masters of a couple of luggers sailing from the port of Yarmouth. John and Henry Thomas (generally known as "Lucky Johnny" and "Gaby") were sons of the noted Breydon gunner, old John Thomas, and both followed the same pursuit on the flats with great success, Johnny gaining his title from oft-repeated luck in securing valuable specimens. Well acquainted with all species of Gulls, Skuas, and other fowl met with off the banks in the North Sea, they invariably carried guns while on the fishing-voyages, and any bird exhibiting an uncommon stage of plumage was sure to attract their attention. When hauling in the vicinity of the sands, if a sudden gale springs up, the crews often experience great risks in saving the nets, and seldom a season passes without several boats with all hands being lost in the broken water, owing to this cause. Many yarns have been spun concerning the desperation with which masters, who were whole or part owners, have hung by their nets till either all hope was lost or the mate or some other of the crew had cut them adrift in time to get up sail and save life. On one occasion, after a voyage in which he had been exposed to the buffetings of a protracted gale, John Thomas gave me a most graphic account of his adventures. Utterly worn out by the continued watching, he had turned in to snatch a few minutes' rest, when roused by one of the crew, who, having sighted a tremendous sea that he imagined must overwhelm them in their partially disabled condition, had rushed breathlessly down, exclaiming "Here we all go together, master!" Though the deck was completely swept, Johnny's lucky star was shining, even if obscured behind a cloud, and having weathered the storm he reached Yarmouth in safety. After landing, however, while undergoing the operation of shaving at a barber's, his well-known features completely hidden by a copious administration of lather, he listened to a marvellous story of the loss of the Gorlestone lugger, of which he was then master, related by a fisherman who declared he had seen her go down with all hands.

Cromer crabs two hundred for one hundred: these, of course, are only wholesale prices. On the Brighton beach, herrings are bought by the salesmen at six score and eight to the hundred, and mackerel at six score and twelve.

None but those who have examined the poisonous concoction would credit the injury inflicted on the fishing population by the vile spirituous liquors sold, bartered, or exchanged for fish, or even gear, by the "Coopers." This infamous trade is carried on by the bum-boats, hailing mostly from Dutch ports, which are perpetually hanging round the fleets of smacks or luggers; unless the skipper resolutely refuses to allow the small boat to be lowered, these harpies frequently dispose of large quantities of their abominable wares, their patrons for the most part being found among the inveterate sots and the younger and more inexperienced of the crews. Drink is doubtless the cause of more disasters among the smacks and luggers than is generally credited, and not unfrequently those who least deserve it suffer. Early in the winter, a few years back, while steaming towards the land, a smack with colours half-mast was sighted in charge of a tug. It was afterwards ascertained that the whole of the hands, with the exception of the master, had been the worse for drink, though the vessel was scarcely an hour out of port. The skipper had apparently stretched over the quarter to knock the trawl-head on to the beam, and having lost his balance had fallen into the water; the crew being helpless, and no vessels sufficiently near to render assistance, the unfortunate man was drowned.

During rough and stormy weather in autumn and winter these birds make their way long distances inland; in East Lothian I often noticed immense flocks settled out in the centre of the largest fields. Like Curlews they invariably paid particular attention to the selection of a spot at some distance from a stone wall, being evidently well aware of the chance of concealment to their enemies offered by a "dry stone dyke." A few of the larger species of Gulls visit the Norfolk broads lying near the coast almost daily throughout the year, the numbers greatly increasing towards autumn, when the herring-boats are off the coast. Stormy weather, particularly if from the east, at this season sometimes brings them in thousands to the inland waters; with a westerly wind the birds may be seen sitting in flocks stretching for half a mile or more along the sandy shore near Horsey, sheltered by the high sand-banks from the force of the gale. Some years back I frequently remarked that the flooded marshes along the coast of Sussex attracted large bodies of Gulls in winter; latterly they appear to have deserted these quarters, resting, when the weather is too rough for them to keep at sea, on the mudflats or the wide-spreading shingle-banks. In all the localities referred to, these gatherings were composed of the three larger Gulls, the present species in all instances being the most numerous.

Every season, in April and May, Lesser Black-backed Gulls are to be seen in the Channel in a state of plumage that might easily be mistaken for the full adult dress; a comparison of one of these birds with a specimen procured at the breeding-stations would, however, show that the tint of the deep grey on the back is far darker, the feathers on the back of the head are also scanty and worn, utterly different from the profuse glossy covering on the head of an adult. That these birds are plentiful off the south coast may be judged from the fact that when throwing overboard fish-liver to attract the Stormy Petrel, as well as when driving the birds from the shoals of mackerel, I have shot from a dozen to a score repeatedly, all in precisely the same state of plumage. These are undoubtedly in the last stage before the perfect mature dress is put on.

From the changes exhibited by this species in captivity, as well as from the observations taken on those in a wild state, I should imagine that the age of five years was invariably reached before the perfect mature plumage was assumed and the bird paired and nested. Though a statement occurs, in the recently published edition of Yarrell, to the effect that the Lesser Black-backed Gull arrives at maturity and breeds when four years old, I prefer to retain my own opinion on the subject, fully believing that these birds are in their sixth year (that is, five years old) before they become perfectly adult.

During their first autumn and winter, juveniles of the Lesser Black-backed and Herring-Gull are much alike in the colouring of their plumage, while the tints of iris, beaks, legs, and webs correspond exactly. The general tone of the plumage of the former is, however, decidedly darker, while youngsters of the latter species are, according to my own experience, invariably of greater size and weight.

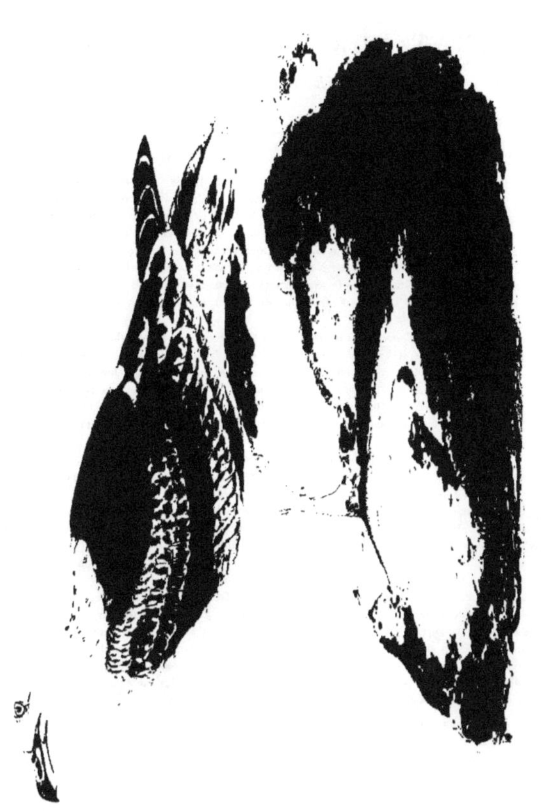

GREAT BLACK-BACKED GULL.

LARUS MARINUS.

This magnificent Gull, the largest of the family that nest on the British Islands, may be met with at all seasons round our coasts. The only breeding-stations I have examined are in the northern counties of the Highlands and on the adjacent islands. The Bass Rock was formerly stated to be frequented during the nesting-season; but this resort must, according to my own observations, have been deserted for nearly five-and-twenty years. Adults and immature in various stages, the latter greatly predominating, are, however, to be seen during the fishing-season in immense numbers in the North Sea as far south as Yarmouth or Lowestoft. Straggling parties and single birds show themselves in the Channel and along the estuaries in the southern and western counties, though in these parts I have never fallen in with them in anything approaching the numbers at times found off the east coast.

The scientific name of *Larus marinus* bestowed on this species would lead those unacquainted with its habits to suppose that it is at least as much addicted to a life on the ocean wave as the Kittiwake and other true sea-fowl. Such, however, is not the case, as many of these fine Gulls breed far inland on small islets in the freshwater lochs and even on the open moors. In several of the northern counties I have known them to frequent the lochs and hill-sides during the whole of summer and early autumn. Towards the end of July 1868, I remarked several small parties, numbering from three or four to a dozen adults, resorting to the small islands on Loch Craigil and other pieces of water in the east of Sutherland. As these birds were repeatedly observed in the same neighbourhood, it is probable they had been robbed of their eggs or young; which would account for their gathering in company at such an early date. On the flooded marshes in Norfolk, Suffolk, and Sussex I have also in winter met with those birds in large parties.

In the Highlands the Great Black-backed Gull causes considerable loss to many of the small sheep-farmers and crofters, who are unable to give the necessary care and protection to the few animals they possess. A weakly ewe is no sooner discovered than she is set upon, and after being either forced into some crevice among the rocks, or slowly butchered by thrusts from their powerful bills, the lamb next falls an easy victim. Such facts, I am aware, have been denied by some writers; but during the last few years several instances have come under my observation, in addition to the reports I have heard from shepherds and small owners. The young of Grouse and many other birds breeding on the moors are also greedily devoured by these robbers, and no exposed egg is safe if once it has attracted their notice. On many occasions I have seen these birds captured in vermin-traps set on the moors, the baits by which they were attracted having been dead sheep, lambs, hares, or fish *. Those that breed along the sea-cliffs seldom make foraging excursions inland, their prey being gathered for the most part along the shore or out at sea; in some instances the nests of their neighbours are also plundered. Though their plumage is pure and spotless as the driven snow, these voracious birds

* I frequently procured large ling of 8 or 10 lbs. from the fishermen on the west coast, and found them excellent bait for all kinds of vermin.

are decidedly omnivorous; carrion, however foul, putrid fish, or any floating refuse, comes by no means amiss when more tempting prey is scarce. A dead seal or a drowned sheep cast up on the shore of a Highland loch frequently draws a large and mixed party to join in the savoury banquet; and it is seldom the Great Black-backed Gull is absent from such gatherings. Along the southern and eastern coasts I have repeatedly watched, during autumn and winter, numbers of this species (for the most part in immature plumage) flying in circles over the tide running from some harbour-mouth, or even the outfall of a large drain, contending at times with Crows and other scavengers for the most attractive morsels.

Any one who has spent much time punt-gunning on the Highland firths along the north-east coast during the first quarter of the year (in autumn and early winter the Gulls have either not come down from the hills or are as yet unaccustomed to this manner of procuring their food) must have been much annoyed by the presence of these Black-backs. No sooner does the gunner make preparations for approaching an unsuspecting bunch of fowl, than three or four screaming Gulls gather over the flock, and after flying round for a time, their excitement increases as the punt draws near, till at last, darting down open-mouthed, they drive every bird from the water and put an end to all chance of a shot. These crafty Gulls are well aware that a meal will easily be secured after the discharge, since cripples are sure to escape from the gunner, only to fall victims to their powerful beaks. Their hunger, however, urges them to attempt the capture before the shot has been fired, and in nine cases out of ten, by these means, they defeat their own ends and ruin the punter's sport. Should these pests by any chance happen to be absent till the cripple-chase has commenced, they often drive birds back to the punt. I have seen numbers of wounded Mallard and Wigeon so terrified by the attacks of the Gulls that the poor birds have taken wing and pitched down within ten or fifteen yards of the punt. In some districts, where there has been much shooting, it is useless to attempt to get near fowl by daylight, however tame they may be, owing to the voracity of the Gulls. In my notes for 1869, while shooting on the Durnoch Firth, I find that, on March 15, at least half a dozen fair shots at bunches of Pintails, all fine drakes, were lost by the Gulls constantly keeping in attendance. When at last it was obvious that every chance would be spoiled, I turned my whole attention to the Gulls, and with punt and shoulder-gun succeeded in administering such a lesson that the few survivors gave the punt a wide berth for the next few days. With poison carefully inserted in certain portions of a dead fowl they might easily have been thinned down; but an unfortunate mishap in the district, resulting from this plan, set me against attempting to clear off the Gulls by these means. Small birds such as Golden Plover or Dunlins, when falling on the water or soft mud, are often carried off by these robbers before they can be retrieved.

Though Great Black-backs are frequently seen during winter on the large broads in the eastern counties, and many wounded birds often fall to their share, I have never found they interfered in the slightest degree with the sport, keeping usually at a respectful distance, and simply picking up any cripples that succeeded in making their escape in the first instance. On January 1, 1873, I captured on Hickley Broad a fine adult in winter plumage in a vermin-trap set for the benefit of some Grey Crows that had been perpetually carrying off the Indian corn and other food supplied to my decoy-Ducks. The trap was placed a few yards outside the Ducks' enclosure, a dead Pochard drake being employed as bait. The following description of this specimen is extracted from my notes:—"Head slightly speckled with grey feathers. Eye silvery white, with vermilion circle. Beak bright chrome-yellow; dark orange spot on lower mandible with purple mark in centre. Interior of mouth pale salmon tint. Legs and feet a pale flesh."

Game-preserving and the increased number of tourists visiting all the most easily accessible quarters have during the past twenty years thinned down the number of these fine birds in many parts of the Highlands, traps clearing off the adults, while the eggs and downy young fall prizes to the wandering collector. For my own part, I have to plead guilty to causing the death of but a couple of broods of young—a pair full-fledged on Loch Shin, and three in the down, a few days old, taken from a nest on an island in Lock Skeanaskaig, in the Coigach district. Though Loch Shin is by no means the "huge ditch" it was described by an old writer *, the scenery is far less wild and romantic than that to be met with a few miles further west. In this locality the Great Black-back used in former days to be numerous; there were, however, but four or five pairs flying over the water when I was in search of juveniles of this species in 1868. The following short and condensed extract from my notes will give some idea as to the present scarcity of this Gull:—

"June 30. Taking the india-rubber boat on the wagonette, left Lairg early and drove along the north side of Loch Shin within a mile or two of Overskaig. For the first few miles Common Gulls and Lesser Black-backs alone were seen, and it was not till reaching a spot off Fiag Island that any of the Larger Black-backs came in view. Leaving the trap on the road, I made my way to the water-side and watched, for at least an hour, the action of a couple of pairs and a single bird or two. After hovering round the islands for some time, the whole party took their departure towards the east, and I could obtain no evidence that they had either eggs or young in the immediate vicinity. After driving a mile or two further west and meeting with no signs of the birds of which I was in quest, we selected a spot where the road ran within a short distance of the loch-side, and having blown out the boat, I embarked in order to thoroughly examine the islands and then pull back to Lairg. On approaching Fiag, the larger of the two islands, a pair of Gulls at once came in view, circling round with angry screams, greatly enraged at the intrusion on their haunts. The adventure on Loch Maree a few weeks previous (recorded under the heading of the Osprey) having by no means faded from my memory, I first selected a sheltered spot where the boat would be secure from any sudden gust of wind, and next closely searched the patches of long grass on the south side. A nest evidently new, but empty, was soon discovered, most cunningly concealed, considering its size, among the roots and coarse herbage in a clump of small straggling bushes that stretched down towards the shore. Fresh footsteps, however, in the soft gravel, plainly indicated that the spot had been lately visited, and the contents most probably carried off. Herons were breeding in several of the low stunted trees, but no other nests were visible, though two or three pairs of Grey Lag Geese showed considerable disinclination to quit that part of the loch. Nothing appearing on the smaller island, I started before a favourable breeze for Lairg, the pair of Gulls still keeping company, sailing slowly round and round, giving vent at times to a melancholy cry. As the wind freshened, my work was easy, good progress being made with little more exertion than simply keeping the boat before the swell, another pair of Gulls soon joining my former attendants, their excessive consternation as I drew further east plainly indicating the presence of young at no great distance. While passing close to the north of a small sandy islet, the highest point of which was little more than a foot or two above the surface of the loch, I caught sight of a couple of young Gulls running in a crouching attitude down to the water's edge and at once striking out for the shore. Turning the boat under the shelter of the island, I soon worked to windward of the two birds, who continued swimming side by side, though making but little headway against the cross swell. Drawing rapidly upon them, I succeeded in the first attempt in clutching the pair, and the larger seizing his smaller brother, the two were comfortably stowed in the bottom of the boat beneath the seat before they had time to offer the slightest resistance. The male and female were now flying round in the greatest excitement, and being anxious to secure the entire family as specimens, I took the first chance and dropped one of the pair as they swooped past the boat. A minute later the survivor again sailed over at a considerable height, and being stopped dead as a stone in the air, doubled up and dropped within a foot of the boat, so close, in fact, that a wing brushed the side in the act

* Macculloch.

of falling. Though not the slightest damage was inflicted, I felt doubtful for a moment as to what might have been the result of the concussion had the dead body struck the inflated cylinder of the boat. The bird wounded with the first barrel was washing along a short distance to leeward and was soon picked up. Having now obtained both adults and young, a few minutes were spent in a vain attempt to arrange the cargo in a satisfactory manner before continuing my voyage. It was utterly impossible in my present position to kill the old bird, and the young, though quiet and sulky for the moment, were somewhat snappish. Being, moreover, uninjured as well as scarcely less powerful than their parents, I anticipated an outbreak, and luckily pulled for a point of sand on the north shore. But half the distance had been passed over when, owing to the cross swell, the spray broke freely over the gunwale and fell on the captives in just sufficient quantities to rouse up the wounded bird, which instantly commenced a furious attack on whatever it could lay hold of. The floor of an india-rubber boat is necessarily confined, and the example of their parent probably exciting the juveniles, a general mutiny broke out. The narrow seat afforded but slight protection, and, though luckily never adopting the true Highland costume, I suffered severely from the sharp beaks of the whole party, my cramped position and the size of the craft rendering any effectual resistance impossible *. At last the shore was reached, and assistance at once procured, as the shots had been witnessed from the wagonette. After being relieved of my troublesome passengers, whom I despatched by road, far better progress was made, and the landing-place in front of the inn at Lairg was reached in three and a half hours after going afloat fifteen miles up the loch—not such bad travelling, considering the delay in searching the islands and securing the Gulls; a favourable breeze, quite as much as was agreeable, had, however, contributed in no small degree to the speed. The beaks of the young were of a dark horn tint, light at the points and flesh-coloured at the base of the lower mandible; inside of mouth pale flesh; iris hazel; legs and feet pale livid flesh."

The nestlings while in the down are of a pale brown or sandy colour, with a few black marks on the back of the head and neck, the remainder of the down being here and there speckled with grey. The general tint is somewhat paler than the down on the young of the Lesser Black-back. The beak in the early stage is very pale horn, light at the tip, flesh at the base; iris hazel; legs and feet pale livid flesh. The beak rapidly alters, changing to a darker horn as the birds grow older and the feathers are assumed. The eyes do not change during the first year. The young in the downy stage I examined in two or three nests on the islands in Loch Skennaskaig on June 13, 1878, one brood being secured as specimens. Numbers of adults are seen in summer about this magnificent sheet of water, and several pairs breed on the smaller rocky islets and on the wet and boggy parts of the adjoining moors. The views from the summit of the larger island, near the centre of the loch, are perhaps unsurpassed in the Highlands. An immense stretch of wild and rugged moorland, backed by many lofty mountains, whose strange glassy slopes and steep broken precipices meet the gaze in almost every direction, form a panorama that can scarce fail to strike the beholder. A stone dyke has been built on the top of Suilven, to prevent the sheep approaching a dangerous part of the hill, and this structure in clear weather is faintly visible, otherwise there is not a sign of human intervention with the primitive condition of the country over all the wide expanse.

There is little or no change in the plumage of this species during the second year. At the age of twelve months the immature bird has but increased in bulk, the iris, beak, and legs being much the same as during the first autumn. While in this stage, unless worn out by protracted gales, they seldom stray far inland, the shores of the salt-water lochs and the sea-coast being their usual haunts. The rocky islands off the Hebrides and along the western mainland are at times densely covered by numbers of immature Gulls of various species. In the interior of the Long Island the upper waters of Seaforth Loch, above the narrows, are frequently resorted

* The bite of the Great Black-back is sharper than that of any other of our sea-fowl. On one occasion the shore-master at Tain (an official who has charge of the mussel-banks) seized hold of a bird I had winged, and a deep cut on each side of the thumb was inflicted, entirely incapacitating the hand for some weeks.

to by many of the larger Gulls. Here I procured several specimens of this species in various stages during the early part of May 1877. Throughout this remote district the whole of the sea-fowl appeared perfectly fearless, affording excellent opportunities for selecting those stages of which I was in search without committing useless slaughter.

During my wanderings in the Lews, I invariably found the inhabitants a most obliging race of people; from the highest to the lowest they appeared eager to assist a stranger by all means in their power. One native, however, was met with whose anxiety to please was carried to such an extent that even the most exacting individual must have been thoroughly satisfied by his endeavours. While returning from Loch Seaforth Head one evening, accompanied by a couple of keepers, after obtaining a few specimens of both the Greater and Lesser Black-backs in intermediate stages of plumage, we were overtaken on the road towards the Lodge by a light cart employed to procure supplies from Stornoway. As three or four miles had yet to be passed over, I stopped the man and signified my intention of driving the remainder of the way. The birds were carefully packed in the back of the cart, and the main part of the instructions to the driver having been given by the keepers in the native tongue, I did not discover till we were some distance on our way that my companion possessed but an exceedingly slight knowledge of the English language. The last mile or so towards the Lodge was a steep descent, the road being cut in some parts into the bare face of the hill-side. As the track was narrow, precipitous rocks on the one side and a rough and stony valley on the other, I considered that the pace at which we had been travelling might with advantage be reduced, and the willing beast kept in hand downhill. On attempting to communicate my ideas on the subject, a nod, a grunt, and a grin were the sole response, the reins were dropped on the animal's back, and several well-intentioned cuts with the whip having been administered, we dashed downhill at a pace which, though decidedly risky, promised, if all went well, to bring us speedily to our journey's end. Once off on our wild career, it was useless to interfere, so I quietly awaited the course of events with, I must confess, considerable doubt as to the result. Though his manner was peculiar, the driver proved himself a more efficient whip than I had anticipated, and eventually pulled up in the yard of the Lodge in a style that would have done credit to the jauntiest Hansom cabby in St. James's Street. He was evidently highly pleased with his performance, and had just sufficient English to return profuse thanks for a trifle with which I presented him. Being uncertain whether the man had not suddenly taken leave of his senses, so strangely had he laughed and conducted himself while dashing down the hill, I made inquiries shortly after, and learned he had understood that I was desirous of increasing the speed, and had consequently done his best to oblige, concluding by remarking that of all the parties he had ever driven, I was by far the strangest to please, but he was in hopes the pace had given satisfaction.

The first signs of the black feathers on the back of this species do not, according to my own experience, commence to show till the third year. Various changes are then gone through each succeeding season till the sixth year, when the adult plumage is put on. After this is once assumed, the bird undergoes but the periodical change into the winter dress, which simply consists in the pure white of the head and back part of the neck exhibiting a few specks of grey; the colouring of the beak also alters slightly.

In the various intermediate stages, I have obtained specimens on the coast of Norfolk and also on the broads in the interior of the county; they are most commonly met with during winter in these parts.

One particularly handsome specimen (figured) was shot off the island of Soya, a few miles south-west of Loch Inver, on June 10, 1877. By means of a quantity of fish-liver and oil poured on the water about a mile to windward of the island, clouds of Gulls and a score or so of Terns were in the course of a few minutes attracted to the spot, hovering over the surface and dipping down and securing the floating particles of food. The hungry birds, while circling round and contending over their prey, took little notice of the boat. Though this plan was tried on several occasions along different parts of the coast, and hundreds if not thousands of sea-fowl were at times attracted, I could detect no strangers among the swarms collected, and but this single specimen of the Black-back was obtained.

The nest, like that of most sea-birds, is by no means elaborately constructed, differing slightly in its composition according to the locality. On the sea-coast it is placed among the rough herbage on the upper ledge of a rocky cliff, or at times in some cavity among the bare stones. Coarse grass with strands from any adjacent plants are the principal materials, with now and then a small quantity of fine dead sea-weed. A few feathers are usually to be seen on the nest or scattered around, but these are probably plucked by the birds themselves while cleaning their plumage. On the islands in the freshwater lochs and on the open moors the nests are constructed almost entirely of a mixture of fine and coarse grasses, with an occasional blade or two of rush or other marsh-plant. In June 1869 I watched a female sitting on her nest on a grassy ledge in the face of the cliffs near Duncansby Head in Caithness. The steep and narrow ravine, owing to the shade thrown by the lofty rocks, was dark and gloomy in the extreme, the white plumage of the Gull standing out clearly defined and making a most striking picture. Though Kittiwakes, Razorbills, and Herring-Gulls were all breeding at no great distance, the ledge on which the Black-back had taken up its quarters was otherwise untenanted save by a pair of Black Guillemots, whose eggs or young were snugly concealed under a large block of stone lying on the grass within a few feet of the nest of the Gull. When the active little divers pitched on the ledge, and rapidly made their way to their domicile beneath the slab of rock, I noticed the Gull occasionally stretching forward her head, as if warning them against too near an approach.

The eggs of most of the larger Gulls are much alike in colouring, and except by means of a large series of coloured plates it is impossible to give any description that would be of service.

HERRING-GULL.

LARUS ARGENTATUS.

THIS well-known species is to be met with all round the shores of the British Islands; though, as a rule, frequenting the sea-coast and the adjacent islands, large bodies may occasionally be seen inland.

Unless from a supposed similarity in tints and colouring to a fresh-caught herring (a clear blue-grey pervading the back and a silvery white lighting up the underparts), it is somewhat unintelligible why the name of Herring-Gull should have been applied to this species, the Lesser Black-backed Gull being a far more frequent attendant on the fleets of herring-boats off our coasts during autumn, and consequently a much larger consumer of that fish than *Larus argentatus*. The fishermen of the Firth of Forth, however, are under the impression that when the birds are numerous along shore shoals of herrings are approaching the coast. I was unable to ascertain that these fish formed any part of the food of the Gulls resorting to the Firth at that season; indeed it was only whilst they were accompanying the fleets of boats or frequenting the harbours where the cargoes of fish were discharged that I noticed their prey consisted of herrings. Some years ago, while sailing past the Bass Rock towards the close of summer, a flock of perhaps one hundred and fifty Gulls (among which several adults of this species were conspicuous) was observed scattered over the ledges above the landing-place. One of the crew at once drew my attention to these birds, remarking that their presence at this season was a sure sign that herrings were in the Firth. Being anxious to ascertain, if possible, on what they had been feeding, I discharged both barrels of a heavy rifle at the flock, in the expectation that one might be struck, or that, in their alarm at the sound of the bullets, some food might be disgorged. The whole body sprung instantly on wing, and flapping out to sea before we could approach within range of the shoulder-gun, escaped without a bird being obtained. On landing and examining the ledges on which they had been resting, it was discovered that several had thrown up the contents of their stomachs, consisting only of perfectly undigested grain, with a number of diminutive mussel-shells. A few days later (August 22nd, 1874) several immature Herring-Gulls were shot along the coast; but I was again unable to establish the fact that herrings had formed any portion of their diet; at this season these birds appear to procure much of their food inland.

Judging from my own observations, I should be inclined to think that the farmer rather than the game-preserver suffered from the damage caused by this species. In many northern districts complaints have been raised, accusing these birds of attacking the roots of turnips when other food was scarce, also of making considerable inroads on the newly sown grain. During the seasons spent on the moors I have not seen a single individual captured in the vermin-traps set for the destruction of other Gulls, or observed them preying on either young game or eggs. In preference to the Grouse-moors in the north, these birds appear to resort to the cultivated tracts of land in the neighbourhood of the coasts, where, after feeding in large flocks on the fields, they retire to the rocks to rest. It is, however, a mistake to assert that the Herring-Gull is utterly blameless; on the Fern Islands I learned from the men in charge of the egging-business that should the

Eider Ducks leave their nests exposed, these birds, in company with the Lesser Black-backed, would immediately swoop down and destroy the eggs. Before leaving the islands I had ample opportunities of verifying the truth of these statements; on other parts of the coast also charges of a similar nature were brought against this species. On my last visit to the Ferns I was informed by the egg-gatherers that Herring-Gulls had of late years greatly decreased in numbers. As but few pairs now nested on the rocks, the price of their eggs to dealers and collectors was somewhat heavy compared with those of the Lesser Black-backed Gulls, the latter fetching but fourpence a dozen, while the former were sold at threepence each. As far as I could judge, the Herring-Gulls were considered too scarce to be plundered without fear of entirely exterminating the small colony still resorting to their old haunts. To distinguish the eggs of the two species, however, is almost impossible, and doubtless the one did duty for the other, giving equal satisfaction to the majority of purchasers. As previously stated, this species was far less abundant than the Lesser Black-backed Gulls in the large flocks keeping company with the luggers during the herring-fishing off the east coast. Small parties of twenty or even up to three or four times that number of adults might occasionally be seen; but I never noticed them to form even a quarter of the immense gatherings met with during the height of the season. Though it does not need a skilled ornithologist to distinguish the juveniles of the two species when compared side by side, it is utterly impossible to form an opinion as to the identity of birds in that state of plumage in the clouds circling round the boats while the nets are being hauled.

I have frequently remarked during autumn and winter that large flocks of Gulls, composed for the most part of the Lesser Black-backed, with a few of the Greater Black-backed and Herring-Gulls, make their appearance almost daily on Hickling Broad as well as on other of the larger sheets of fresh water in the east of Norfolk. The birds fly in from the sea-coast usually about midday, and after wheeling round the broad for a time, settle near the centre, where some hours are often spent ducking and slushing about in the water before they return to the sea. It is probable that these Gulls having made their morning meal on the coast or out at sea when the nets of the herring-boats were hauled, visit the inland waters to wash and drink. Immense numbers also are attracted to the broads should disease break out among the fish; in the summer of 1875, a few pike and rudd having perished from the effects of some impurity in the water, the Gulls eagerly sought out the decomposing remains floating round the edges *. About twenty years ago the whole of the finny tribe suffered severely from some unknown cause, the mortality being exceedingly heavy for several weeks. Gulls of all descriptions were reported by the natives to have arrived in swarms at that time to feast on the repulsive banquet provided by the rotten and bloated carcasses.

During the autumn of 1871, when shooting on Breydon Water, I noticed quantities of dead eels floating in the tideway along the banks, while others in the last stage of disease were showing themselves on the surface. Immature birds of the two Black-backed and also of the Herring-Gulls were collected in numbers over the river, darting down with loud screams and fighting with the greatest fury over the choicest morsels. The excessively unpleasant condition of the food consumed by these birds may be judged by the fact that two of my puntmen, who had requested that a shot might be taken into the swarm of Gulls in order to secure feathers for bed-making, were utterly unable to make use of twelve or fourteen I turned over. The foul discharge running from the mouths of the birds rendered it impossible to pluck them; and though an attempt was made to remove the feathers with a pair of shears, the unpleasant odour was found to be unbearable. I am entirely ignorant as to the cause or nature of the disease from which the eels suffered; they, however, could often be seen

* In this case there was no disease, the fish having been merely affected by some impurity in the water; the majority succeeded in making their escape by crowding up the dykes to avoid the flow from the main channel, those only that were shut into some remote corner succumbing to the effects of the poison. If taken out while sickly from the broad or river and placed on fish-trunks in the marsh-dykes they rapidly recovered.

protruding their heads above the surface of the water, the throat and a portion of the belly being red and swollen *.

The Jackdaws having proved exceedingly mischievous some years back on the Bass, the tenant of the rock placed bread and butter well seasoned with poison on the ledges to which they resorted. This proceeding certainly had the desired effect, as it cleared off the Daws; the larger Gulls, however, also suffered so heavily that since that date but two or three nests of either the Herring- or Lesser Black-backed Gull have been met with on the grassy slopes on the Bass, formerly a well-frequented breeding-ground. The situation selected for nesting-purposes by this species renders their eggs extremely liable to fall a prey to unscrupulous visitors who explore the upper portions of the rock; consequently it is unlikely that these Gulls will again become numerous in this locality. Some years back I was a witness to the prompt chastisement administered to an egg-stealer endeavouring to make his way off the rocks with pockets distended by plunder. A custom in those days prevailed among the farmers in the district to give a holiday annually to their labourers, when the whole party were conveyed for an outing to the Bass; on such occasions extra precautions were necessary in order to prevent the nests on the higher ledges of the rock being robbed of their contents. Three or four of the boatmen from Canty Bay were usually stationed at the upper part of the old fortifications in order to put a stop to any infringement of the regulations; and on this occasion easily detecting the large number of eggs that an overgrown laddie was attempting to carry off, the culprit, having previously protested his innocence, was tripped up and rolled for a time on the grass. Oh! what a mess that poor laddie was in! It is highly improbable that the Gulls' eggs suffered from his attentions on the occasion of his next visit to the rock.

On the shoal water that stretches at low tide for many miles round some of the dangerous outlying sand-banks off our eastern coasts, I have remarked large bodies of these Gulls, together with the Larger and the Lesser Black-backed Gulls, busily engaged in darting down open-mouthed to the waves, plunging their heads below the surface and apparently endeavouring to capture fish. The distance at which the birds were watched was invariably too great to clearly ascertain the nature of their prey, which was evidently extremely abundant. Having had my attention drawn to the wounds seen on numbers of codling taken in Yarmouth harbour †, I came to the conclusion that the cuts could only have been inflicted by the sharp beak of some bird, and in all probability these were the fish of which the Gulls were in pursuit. Out of ten score codling captured by the hook on one occasion, about twenty exhibited deep cuts on the back and sides. Having repeatedly examined the injuries inflicted on fish when fixed or bitten by congers, cuttle, or their larger relatives, I noticed that these wounds were totally different, being clean cuts, as if sliced with a knife—just such, in fact, as the sharp and powerful bills of Gulls would cause when, having seized the fish too near the tail, their slippery prey succeeded in escaping. The codling exhibiting these marks were of about the same length and bulk as an ordinary mackerel; judging from the capabilities of Gulls kept in confinement, I should think that a fish of this size is about the utmost these birds can swallow. The skipper of one of the luggers to whom I showed some of these fish was under the impression that the wounds must have been caused by the broken bolts in the old piles or wreckage about the harbour; others thought the paddles of the tugs might be responsible; considering, however, that the whole of the cuts were either perfectly straight up and down or at a very slight angle, they could hardly have been caused by these means.

* For those who indulge in such luxuries as eel-pies it may not be out of place to mention that, as far as I was able to ascertain, there was no diminution in the quantity of eels sent off to the London market in consequence of the disease.

† Formerly when the luggers brought up in the roads and the cargoes of herring were ferried ashore, the codling which arrived off the coast at that season did not enter the harbour in any numbers. Since the construction of the new fish-wharf at the river-side they have deserted their old quarters in the roads and make their way up the harbour in immense shoals, the quantity of broken fish and other refuse carried down by the tide having evidently attracted them to the spot.

The full-plumaged adult of *Larus argentatus* is an extremely showy bird if viewed sweeping along the shore when dark and threatening clouds obscure the wintry sun. Hundreds of Gulls in the immature state may be fishing over the stormy waves without being regarded with more than a passing glance, though the snowy breast and well-defined colours on the back and head of the old birds of this species are sure to attract attention. On many parts of the coast large flocks of Herring-Gulls in company with other sea-fowl may be seen gathered off harbour-mouths eagerly waiting the ebb-tide, when the refuse carried out to sea affords occupation to the noisy and quarrelsome assemblage for several hours.

Having kept Herring-Gulls in confinement for many years, I conclude that they do not assume the perfectly adult dress till the fifth or sixth year. Though statements to the effect that these Gulls have been seen breeding while still exhibiting the immature plumage appear in print, I have entirely failed to note this fact. Great difference may at times be observed in the size and weight of individuals of this species, though they seldom vary in this respect so much as the Greater Black-backed Gull. The clouded markings on the heads of adults in winter are usually but lightly diffused over the feathers of the crown and back of the neck; a few individuals, however, whose appearance is exceedingly striking may occasionally be met with at this season. On the 24th of January, 1873, I procured on Hickling Broad a specimen to which my attention was attracted by the dark colouring of the head. On examining the bird, the whole of the head proved to be thickly speckled and streaked with black and grey, terminating in a perfect and well-defined ring round the neck, the remainder of the plumage being similar to that usually exhibited in winter. Iris pale yellow, with pale orange circle. Beak and legs as in summer. So early as November 2nd, according to my notes, adults have been observed showing the winter dress.

Many pages and numbers of coloured plates would be needed to give an accurate description of the various immature stages of the Herring-Gull. The shade of the iris is gradually transformed from a dark hazel in the nestling to a pale silvery yellow in the adult. The beak also, which is of a pale flesh tint with clouded point in the downy juvenile, changes by degrees till the bright yellow with blood-red blotch on the lower mandible is reached; this, when once assumed, does not alter with the seasons. An immature bird (probably between two and three years old) shot near the Bass on the 22nd of August, 1874, may be described as follows:—Head, back of neck, and throat lightly speckled with grey, darker over the ear-coverts. Back clear blue. The wings partially speckled, a small space of clear blue showing in the centre, the markings on the shoulders being darkest; primaries black, though one or two feathers exhibited a slight edging of white. Tail clouded with grey, the side feathers darkest. Iris pale drab. Beak pale livid flesh at base, dusky horn towards point, with slight indication of orange on upper mandible. Legs and feet pale livid flesh tint. In the earlier stages the beak is of a livid blue flesh tint at the base, and dark horn towards the point, which, again, is lighter.

The knowledge concerning the changes of plumage undergone by several species of Gulls appears to have been exceedingly limited, even so late as the days of Bewick. We learn from the writings of the celebrated wood-engraver that considerable doubt existed as to whether the Black-backed Gull, the Herring-Gull, the Wagel, and even the Glaucous Gull were not merely different stages of one and the same species.

Though I failed to ascertain, from observations made on this species in a wild state, that they are given to destroying the young of other birds or those less powerful than themselves, this habit is certainly acquired in confinement. Three Gulls that I kept for many years made a meal off a couple of young Blackbirds that escaped from a cage in which they were being reared. A Water-Rail, with which they had lived in amity for some time, was eventually devoured; and I discovered them in the very act of murdering an unfortunate Norfolk Plover which had been their companion for five or six weeks. When I happened to look over into their enclosure, two of them were holding down the wretched bird, while a third was hammering at its skull with repeated strokes from its powerful bill; the whole party immediately retreated on my appearance, but the deed was done. It could not have been hunger that caused them to commit these barbarities, as, in addition to the fish with which they were regularly furnished, they had at all times either a pan of soaked bread or a supply of corn, on which I have frequently observed them to feed when in a wild state. This bread and corn was a great attraction to the Sparrows, which were not unfrequently snapped up and immediately swallowed; a mouse or a young rat that was let loose in their enclosure was also speedily captured and gulped down.

A short account of the singular behaviour of these Gulls may not be out of place, as many extraordinary antics, never noticed in a wild state, were occasionally indulged in. After keeping the birds for some years in an enclosure laid down with turf, I was at length forced to banish them to other quarters, as they persisted in tearing up the grass by the roots. To one of their number, whose vocal powers were most astounding, the name of Sims Reeves had been given, while two younger birds were known as Moody and Sankey *. After several years had been spent in confinement, I remarked that Sims Reeves asserted his authority over the other two in the most overbearing manner, driving them round and round the pond, the two poor wretches meekly trotting in front of him, while he every now and then gave vent to the most melancholy and piercing screams. Ascertaining at length that they would not live peaceably together, Sims Reeves was allowed to go with his wing unclipped, and in due course took his departure. No sooner had he gone than Moody at once became "boss," and the last state of poor Sankey was no better than the first. At times they were quiet and contented enough; resting side by side on the grass, they appeared the best of friends. Without the slightest warning, however, Moody would arise, and when he had cleared his throat by a preliminary "caterwaul," the submissive Sankey, having learned by experience that it would not do to be caught, would be up and off. Then, with his head drawn back between his shoulders and his feathers slightly puffed out, Moody would follow in his wake. For an hour or so this mournful procession round and round the pond would continue. At last Moody would stop, Sankey also pulling up at the distance of a yard or two, Moody leading; they would then commence a duet *à la tom-cat*, when suddenly dropping on their breasts on the ground, they would turn rapidly round several times, and at last attack the grass in the most excited manner, tearing it up by the roots and scattering the fragments in every direction. This proceeding was accompanied by the most melancholy cries and screams; and when it is stated that the voice of Grimalkin in his happiest or, rather, his unhappiest mood is almost sweet and pleasing to the ear, compared with the discordant wailing of these infatuated birds, one may judge of the nature of their performance. Whether these antics were intended for courtship or defiance I was utterly at a loss to comprehend.

* I found it necessary to give names to each and all of the birds kept in confinement, so that directions might, if required, be sent for special treatment to any particular individual during my absence from home.

GLAUCOUS GULL.

LARUS GLAUCUS.

THE Glaucous Gull is said by several writers to be most numerous off the Shetland and Orkney Islands and in the North Sea; small parties and large flocks of birds, in various stages of plumage, are also seen at times frequenting the seas around the British Islands.

My own experience concerning this species, gained from personal observation, is unfortunately somewhat scanty. While staying at Yarmouth during the winter months, I now and then saw a few immature birds in the Roads and off the harbour-mouth; in the terrible November gale of 1872, that continued for six days, commencing on the 11th, I noticed three or four, in immature plumage, flying about along the shore, but failed to obtain a shot, as they all kept too far out to sea. In somewhat finer weather on the 30th of October, 1879, I fired a shot from the beach near the head of the north pier, at a young bird, and it fell nearly a quarter of a mile out at sea in the Roads. Then, driving across the wharf by the river, I happened to find the tugboat required, and steamed out of the harbour and thoroughly searched the Roads, but failed to discover any signs of the bird. Many years ago, while shooting in the marshes near Bulverhithe, on the Sussex coast, between St. Leonards-on-Sea and Bexhill, I obtained a capital view of a magnificent adult in perfect plumage; the bird, however, unfortunately passed on out of range, flying west a short distance inside the shingle-banks. The following day a specimen exactly corresponding in every respect, and perhaps the same, as I made a careful examination, was brought into a bird-stuffer's shop at St. Leonards-on-Sea; this fine bird was reported to have been shot on the broad expanse of shingle that stretches along the shore of the Channel adjoining Pevensey Level.

The only Glaucous Gull I have so far procured was obtained near Hickling Broad on the 27th of November, 1874. While proceeding along the marsh-wall from Heigham Sounds towards Hickling Broad, I caught sight of the bird flying along over the course of Deep Dyke, the river between the Broad and the Sounds, and immediately recognized an immature Glaucous. It was evidently making its way inland for a drink and a wash in the fresh water, after having left the open sea about mid-day, as hundreds of the larger Gulls are accustomed to do at this time of year and also earlier in the season, when the large fleet of herring-boats is fishing off the coast. As the distance was fully sixty or seventy yards, it was doubtful whether a charge of shot from a shoulder-gun would have much effect. Having, however, sufficient time to change the cartridges loaded with No. 3 shot, with which my heavy 10-bore was charged, for some with No. 1, I let him have both barrels, just as he came in line with us, flapping slowly up the course of the river. It was evident in a moment that the dose I gave him had taken effect; wavering for a second or two, he fell away towards the north, passing close over the keeper's house on Whiteslea Island, and finally rising with his last effort a few yards in the air, fell headlong into the marshes. As my boats and punts were following, the men were hailed, and the deep water in the dyke and the small broad at Whiteslea were soon crossed, and landing at the stage by the house we passed out by the bridge on to the marshes beyond, and keeping the line taken, soon detected the bird lying most conspicuously with its wings spread out on a large bed of rushes. It proved to be an immature bird, probably in its first year, the plumage being rather more darkly marked than is usually seen. The colours of the soft parts were as follows:—Upper and lower mandibles a pale pink flesh; points dark horn, almost black, though a lighter tint showed towards the extremity; legs, toes, and webs a dirty flesh, and nails a pale brown; iris dark hazel.

While knocking about with the herring-fleet in the North Sea, I noticed three or four immature birds, in company with the swarms of large Gulls, about twenty or thirty miles off the land outside the Cross Sands; they, however, invariably managed to keep out of range. While sailing off the harbour at Yarmouth, on the 31st of October, 1872, in one of the beach yawls, we observed a fine mature Glaucous in perfect adult plumage; the bird appeared shy, for although we followed it for some time, and afterwards waited for an hour or more in the stream of the harbour tide, where the Gulls were feeding, we had no chance of getting within range for a shot.

The men who look after my punts and boats at Shoreham sent word into Brighton on the 14th of November, 1883, to inform us that a pair of these birds, in adult plumage, were feeding and keeping company with a large flock of Gulls along the shore near Lancing. It was too late to start when the message arrived, and on the following morning no signs of the strangers could be detected, though we were afloat by daylight, and proceeded several miles along the coast towards the west, where the Gulls where then harbouring.

While punt-gunning in the Dornoch Firth, off Morangie, Tain, and Golspie, in the winter of 1868, I noticed several of this species, apparently all in immature stages, a few showing signs of a change and probably more advanced towards maturity. These proved almost as great a nuisance as the Great Black-backed Gulls; they carried off several Plovers that had been knocked down and run beyond the range of the shoulder-gun, and also repeatedly put up the Ducks while we were sculling to them floating quietly on the Firth, utterly unsuspicious of danger. This species proved far more wary than their relations the Black-backed Gulls, which had been mostly killed down, and they consequently escaped being shot.

If I am not mistaken, there was some years ago a letter in the 'Field,' in which a traveller, who had visited the Arctic Regions, described a romantic attachment said to exist between the Glaucous Gull and the seal, the Gull being supposed to warn his friend the seal of impending danger; I am afraid, however, if the truth was known, that it was only a very base kind of cupboard love. The Gull having a pleasing remembrance of savoury chunks of seal's blubber, which may have fallen to his share at some previous time, was simply and solely desirous of eating him, though his greediness was such that he could not wait till the fatal bullet and knife had done their work, but must needs commence the attack, without assistance, at the first opportunity. This is my own opinion on the subject, judging from the behaviour of the larger Gulls in driving up the flocks of wildfowl when they detect the punts setting up to them; well aware that when a shot is fired they may have a chance of securing prey, they dash forward to seize their intended victims before the gunner is within range. Many a shot did I lose on the Dornoch Firth, till the water had been cleared of all these voracious birds.

KITTIWAKE.

LARUS TRIDACTYLUS.

None of our British birds have been so horribly persecuted as the unfortunate Kittiwake; luckily many of the breeding-stations to which this species resorts are too remote to suffer from the depredations inflicted on those within reach of shooters and the dealers profiting by the sale of the birds when procured. I have had many opportunities for watching the slaughter of this species in several parts of the country, and also ascertained that immense numbers had been obtained in the Channel off Brighton, their wings in due course passing into the possession of the plumassiers. The manner in which this species has been destroyed and, in some localities, almost cleared off is described by several writers, and I have selected a few extracts that afford a better insight into the barbarity to which they have been exposed than I could give from what has come under my own observation. In the fourth edition of Yarrell's 'History of British Birds,' revised and enlarged by Howard Saunders, F.L.S., F.Z.S., the following remarks are to be found under the heading of the Kittiwake Gull:—

"Some years ago, when the plumes of birds were much worn in ladies' hats—a fashion which any season may be revived—the barred wing of the young Kittiwake was in great demand for this purpose, and vast numbers were slaughtered at their breeding-haunts. At Clovelly, opposite Lundy Island, there was a regular staff for preparing the plumes, and fishing-smacks with extra boats and crews used to commence their work of destruction at Lundy Island by daybreak on the 1st of August, continuing this proceeding for upwards of a fortnight. In many cases the wings were torn off the wounded birds before they were dead, the mangled victims being tossed back into the water; and the Editor has seen hundreds of young birds dead, or dying of starvation in the nests, through want of their parents' care, for in the heat of the fusillade no distinction was made between old and young. On one day 700 birds were sent back to Clovelly, on another 500, and so on: and, allowing for the starved nestlings, it is well within the mark to say that at least 9000 of these inoffensive birds were destroyed during the fortnight."

William MacGillivray, in his 'History of British Birds,' makes the following remarks, which prove that the same senseless destruction of this species was also carried on many years ago in the south of Scotland:—

"Human nature is so perverse that reason affords but a feeble check to appetite and impulse, else I should here deprecate the useless slaughter of these innocent birds. Parties are formed on our eastern coast for the sole purpose of shooting them: and I have seen a person station himself on the top of the Kittiwake cliff of the Isle of May, and shoot incessantly for several hours, without so much as afterwards picking up a single individual of the many killed and maimed birds with which the smooth water was strewn beneath. Nay, I have seen one who, in his books, admonishes you, with great solemnity, of the sin of shooting birds of any kind unless for some useful purpose, fire away at the poor Kittiwakes with as much glee as a school-boy at Sparrows. It is, in fact, human nature, tyrannical and most unamiable. The noise of guns does not always frighten the sitting birds from their nests, and those which have left them presently return, when the boat has advanced a short way."

In the 'Natural History of Ireland' by William Thompson, published in 1851, it is stated :—

"At Ballantrae, on the coast of Ayrshire, these birds are commonly taken, in the following manner, by idle boys. They bait hooks with the liver of the cod-fish, and fling them as far out from the shore as possible, having a stone as a counterpoise to the gull's weight attached to the opposite end of the string, and left at the edge of the water. They then retire to such a distance as to allow the victims to come freely to the bait, and so soon as this is swallowed, they hasten to the stone and draw in the line with the hooked gull at its other extremity. Various species of gulls have been thus taken. The Kittiwakes are purchased on the spot at a penny each for the sake of their feathers, and a person of my acquaintance there has obtained as many of them from birds captured in this manner, as have sufficed to stuff some pillows."

I shall conclude my account of the persecution to which this species is subjected by a short extract from Mr. Dresser's 'Birds of Europe' :—

"Next to the Herring-Gull the Kittiwake is the commonest Gull in the south-western counties, and is equally numerous at all times of the year. There is a large breeding-station at Lundy Island, also on some of the granitic cliffs near the Land's End, and at various other places on the southern coasts, the Cob Rock, off Berry Head, affording a home to a small colony, as Lord Lilford has informed me. Great numbers of Gulls follow the shoals of sprats into the muddy bays of the Bristol Channel in the winter season; and at Weston-super-Mare it is a common amusement with boys to place small jins along the shore baited with broken fish; and in this manner numbers of Gulls are easily captured, Kittiwakes and Brown-headed Gulls being most largely represented among the victims."

Dresser informs us that this species satisfies its hunger by partaking of a variety of tiny morsels:—"It feeds on small fishes, crustacea, and other marine animals, which it usually obtains from the surface of the water, over which it hovers with elevated wings when picking up its food." Other authors confirm these statements, but I am of opinion that the birds require more substantial nourishment than what is referred to. In June 1865 I was staying at North Berwick, on the shore of the Firth of Forth, and took the opportunity of securing specimens of all the sea-birds breeding at the Bass Rock. After shooting a couple of Kittiwakes which flew past the boat near the island of Craig Leith, we picked them up and while holding one of them by the leg and shaking the water from its feathers, three large herrings dropped from its mouth. The other, I ascertained later on, had one fish of the same kind with a quantity of some small fry: herrings were at that time very plentiful in the Firth.

I never observed a Kittiwake far away from the sea, though during the terrible gale in November 1872, which continued for six days, a few that had been knocking about in the roads off Yarmouth were carried on to Breydon mudflats by the frightful gusts of wind, and some made their way further inland to the marshes. These, so far as I could ascertain, all returned at the first lull in the storm, and following the course of the river flew slowly out towards the sea to brave once more the gales that sweep over the briny ocean.

William Thompson, in his 'Natural History of Ireland,' gives the following description of the hardships to which these birds are occasionally exposed in severe weather when carried inland by protracted gales, and finally succumbing from exposure to cold and starvation :—

"Isolated instances only of its occurrence in winter, as just indicated, were known to me until 1840, when within the last ten days of January, one old and two young birds were shot in Belfast Bay, and another old bird was found dead; they were mere skeletons, as Kittiwakes procured here at this season have generally been. Only one contained in its stomach any food, which consisted of the remains of several of the crustaceous genus *Idotea*. Between the 20th of February and the 5th of March that year, ten birds, all adult, came under my notice: three shot in Belfast Bay; three found dead on the beach near Holywood, and with them a herring-gull; all seeming to have died a natural death; two were procured at different inland places (one shot and the other found dead), five miles in a direct line from the sea, or if they followed the windings of the river Lagan, nearly double that distance; the two others were obtained near Kirkcubbin, on the borders of Strangford Lough. All these birds were miserably poor in flesh; four of them weighed respectively 10, 9, $8\frac{3}{4}$, and $7\frac{3}{4}$ ounces avoirdupois; Bewick notes the weight of the bird as 14 oz. So light were several of these birds that they were imagined by persons lifting them to be mere skins, put up in a natural form by the taxidermist. In the stomach of one was found a specimen of the fresh-water shell *Paludina impura*; of another, the remains of a crab; one was well filled with earthworms and earth (this bird was killed when 'following the plough'); and the bill of another contained some dry loamy earth; the stomachs of all the others were empty."

I quite agree with MacGillivray when he states:—"With us it is scarcely ever seen inland; nor does it ever search the maritime pastures or the ploughed fields along the shores for worms and larvæ, like the other smaller species of its family. It is an ocean bird, that loves not the haunts of man."

While fishing and shooting in the Channel off Shoreham, Lancing, and Worthing during autumn I remarked that the young Kittiwakes generally put in an appearance soon after the beginning of September. Ample opportunities for observing the habits of this species were met with while steaming in company with the fleets of herring-boats in the North Sea off the sandbanks about Hasborough, Yarmouth, and Lowestoft during the autumns of several years. I noticed they were capable of devouring immense quantities of herrings and any amount of sprats and fish-liver when cut up into small pieces: we used to feed the swarms of these birds that followed the steamboat in order to draw the Skuas in hopes of procuring specimens in some curious stage of plumage. The Kittiwakes would hover in hundreds just over the stern, darting down when small pieces of fish were flung overboard, and seizing the morsel before it reached the water. Occasionally while fishing for cod, of which we frequently captured some weighing between 30 and 35 lbs., outside the Cross Sands, we came across immense shoals of very large silver whiting, and then those terrible pests the dogfish were often attracted around, and all sport was at an end, as they snatched the baits immediately and were hauled up every time the lines were lowered: the only chance then was to steam a few miles off and commence again. The manner in which the crew of the steamboat amused themselves with these ravenous creatures was somewhat amusing. They made a large ball of corks and fastening it on to the tails of about ten or a dozen that were tied together, they flung the lot overboard: prevented by the corks from going down, the dogfish dashed about just below the surface of the water and attracted a large assemblage of Gulls that remained circling over and darting down to see what was to be obtained.

While conversing with the crews of some of the fishing-luggers I learned that the allowance for each of the crew at a meal was thirty-two herrings. Few, of course, were in the habit of consuming what they were entitled to, and I only heard of one man, a native of Potter Heigham in the east of Norfolk, who made the attempt. This strange individual used to hold the fish by the head and tail and tearing off the flesh on the back with his mouth fling the remainder overboard. My informant who has now the charge of my punts on Hickling Broad had, when a "youngker" * in the crew of a fishing-lugger, held the office of cook and was well acquainted with all the hardships to be encountered when afloat in the stormy North Sea. His first attempt to attend to the preparation of the dinner for the crew, he stated, had proved a failure. Having boiled a large cod of about 30 lbs., he took the kettle up to pour the water out, when a heavy sea striking the boat the fish and all the contents went overboard. The accident having been observed by the rest of the crew, no remarks were made, as there was a large stock of cod and herrings in the vessel, and a fresh supply was soon procured and prepared.

It is only two or three years since, when crossing Hickling Broad in a punt, with the man just referred to "quanting," that we detected a fine pike lying quietly in a round hollow in a mass of weeds that covered the bottom. "That's a nice fish," I remarked. "Yes," replied Bob, "we had better have him into the boat." Permission having been given to strike the fish, as I wished to see the operation performed, the punt was slowly worked

* This was the name among the fishing fraternity for the youngest hand in the crew on board the boat.

over the unsuspecting fish, and raising his arm, Bob dashed the point of the quant through its skull, and after a few flaps our prize was secured, and proved, when weighed, to be just over 16 lbs. After taking the pike up to the farmhouse at which we were staying, it was placed in a large dish and conveyed into the larder. During our absence in the afternoon the farmer's wife made an examination of the fish, and on returning she informed us that it had been feeding on white herrings *. This was a strange assertion, and proceeding at once to ascertain the true state of the case, it was obvious at a glance. The pike had evidently snapped up three rud † for his last repast, and decomposition having already set in, and the red off the fins and the scales with their green and orange tinge having disappeared, the fish had assumed the pale blue and white hue which is always exhibited by a fresh-caught herring.

* Fresh-caught herrings are known by this name in the country districts of the east of Norfolk.

† This fish is always spoken of as the "roud" in the Broad district by the natives.

POMATORHINE SKUA.

STERCORARIUS POMATORHINUS.

It is stated by certain authorities that this species has been discovered nesting within the limits of the British Islands. Up to the present time this fact has entirely escaped my notice, and I am only enabled by personal experience to speak of the Pomatorhine Skua as a wanderer along our coasts during autumn and early winter while returning from its summer-quarters in the far north. What course these Skuas follow when working north towards their breeding-stations I have been unable to learn; though constantly at sea for many years during spring and summer off the southern, eastern, and northern coasts, I failed to recognize the species between January and August.

Immense numbers of both old and young annually pass over the North Sea while on their return journey from their summer-haunts. The first-comers may usually be noticed off the south-east coast of Scotland about the middle of August. The earliest arrivals are for the most part, if not entirely, composed of birds exhibiting a state of plumage which I should judge (from the changes of those kept in confinement) to be that preceding the assumption of the perfect adult dress. These for the most part are without the long tail-feathers, though it is evident, in some instances at least, that these appendages have been broken and lost. The few stragglers in perfect plumage that have by this time reached so far south are seldom seen in the vicinity of the shore, keeping for the most part ten, fifteen, or twenty miles off the land; small parties of adults, however, are occasionally observed in the Firth of Forth flying high in the air, making a straight course either due east or west. Having passed altogether three or four autumns on the coast of East Lothian, and spent the greater portion of my time at sea, I have had ample opportunities of noting the arrival of these Skuas and watching their progress towards the south. In this quarter I did not recognize a single specimen of the dark form of the adult in perfect plumage, all that came under my notice being white-breasted or clouded. The adults of this species are easily distinguished on wing from the Arctic Skua when at any height in the air: in these Skuas the white feathers on the underparts of the belly meet the darker shade towards the vent and tail in an almost straight line, and form a conspicuous mark when viewed from below, while in the Arctic the light and dark colours appear to blend gradually together. The flight of the Pomatorhine is also more steady, and while on passage the birds usually keep at a greater elevation.

A few words extracted here and there from my notes, jotted down while at Canty Bay early in the autumn of 1874, will give some idea of the numbers of these Skuas observed in the Firth at that season:—

August 1874. Though constantly at sea during the early part of the month, no Pomatorhines were identified till the 14th (wind north-east and a heavy swell in the Firth), when a small party was noticed flying out to sea halfway between the Bass and the May. On the 19th, weather fine with a light westerly breeze, several small parties of adult Pomatorhines were again seen making their way out to sea; these were all observed about the middle of the Firth. Pomatorhines were in numbers between the Bass and the mainland on the 21st, a few again on the 22nd and 25th, and on the 26th half a dozen were noticed flying together in a straight line (their usual custom), two being fine-plumaged adults, white-breasted with long tails. Several were clouded on the breast and appeared dark; but I could not positively identify a black adult with long tail-feathers. On the 28th, while out watching the small boats from North Berwick, long-lining for haddies, several Pomatorhines were flying round in company with Arctic Skuas, and a specimen or two were obtained. These birds were all in a state of plumage which is apparently the last stage before assuming the mature dress, and none exhibited the elongated tail-feathers, though it is probable that these had been lost. Some long strings of Pomatorhines were noticed on the 29th, flying west. After a heavy gale on the 1st, the weather moderated on the 2nd of September, and it was again possible to get to sea; Pomatorhines were observed at a distance, but no great numbers showed themselves. Some fine adults seen on the 5th; on the 8th Arctic Skuas were in numbers pursuing the Kittiwakes among the islands, principally between Fidra and the Lamb. A few Pomatorhines were also recognized, but they sailed quietly past without heeding the throngs of Gulls below them. The 14th and 18th squally, a few seen outside; 21st and 22nd, blowing a gale from the south-west, both Pomatorhine and Arctic were driven along the coast; they were, however, by no means exhausted, being well able to continue their course. For several days Arctic Skuas were observed, though no Pomatorhines put in an appearance. Blowing a terrible gale from the south-west on October the 2nd and 3rd; numbers of Gulls and Skuas were off in the Bay, but it was too stormy to put to sea, and the drift and haze prevented the glasses being of any service. Weather moderated on the 7th; several Skuas seen in the Firth, outside the islands, a few being evidently Pomatorhine; and still more passed the boat on the 8th, all flying in an easterly direction, high in the air and straight out to sea.

During October and early in November large parties of Pomatorhine Skuas are usually to be found keeping company with the fleet of herring-boats as they work up the coast, procuring their food by attacks on the Gulls congregated round the nets. The numbers that approach the land vary considerably; in fine weather but few are seen, though a gale of wind while they are off the coast at times drives hundreds along the shore, many being met with worn out, starved, and utterly helpless, from the effects of continued rough weather. The appearance of this species after the protracted October gales in 1879 was referred to by several observers as an extraordinary migration, the writers evidently being unaware of the multitudes annually passing over the North Sea, harbouring for a time round our fleets of smacks and luggers and gradually passing south. It occasionally happens that, should the wind continue light, the main bodies keep a line from forty to sixty miles off the east coast. When but few are met with round the larger English boats, I have repeatedly heard of them, and on one occasion observed far greater numbers, resorting to the French craft: many of these are vessels of considerable tonnage, carrying heavy crews, and their fish being in some instances cleaned before stowage, Gulls in countless thousands are in attendance, procuring an abundant supply of food for the ravenous Skuas.

In the autumn of 1872 the weather was not sufficiently rough during October to affect the Skuas; the greater number having passed southerly before the November gales, but few were noticed this season near the land. A short account extracted from my notes will give an idea of the numbers seen off the east coast during October and November.

October.—A few Pomatorhine Skuas in the first plumage were met with on the 7th (fresh breeze south-east) about twenty miles off Yarmouth. Thousands of Lesser Black-backed Gulls, Gannets, and Divers were on the water, between twenty and thirty miles off the land, the Gulls being in large bodies of several hundred in a flock. Two or three Skuas were sitting here and there; occasionally they would rise on wing and chase any of the smaller Gulls that approached, though for the most part they remained as quiet and contented as their neighbours. Their food is evidently procured while the nets are hauled at dawn and during the first hours of daylight, after which they rest and digest their meal, moving occasionally, if put up by a passing boat, with a slow and lazy flight, a short distance over the waves. On the 8th, though the boats nearest the sands had plenty of fish, and numbers of Gulls and Gannets were flying round, no Skuas appeared in sight till after midday, when a few immature Pomatorhines were met with sitting on the water, about twenty miles off the land; they proved to-day especially restless, remaining but a short time in one spot. But one immature bird of the year was observed on the 10th. The nets of a Lowestoft boat were entangled, and as they held a good catch of fish, hundreds of Gulls and Gannets were flying round, appearing at the distance of half a mile like a swarm of bees. As but the single specimen referred to was in attendance on this large gathering, it was obvious that the main body of the Skuas were further off the land. The boats were hauling only a few miles outside the Cross Sands on the 14th, and even the Gulls and Gannets showed in but small numbers, Skuas of all descriptions, with the exception of an adult Arctic, being conspicuous by their absence. On the 28th, though we steamed thirty miles out to sea in a south-easterly direction, and remained in the vicinity of the boats all day, no Skuas were observed. In the evening I met the master of one of the fishing-luggers, who had been at sea nine days, during which time he had only seen one Skua; his information was perfectly reliable, the skipper being an old puntsman who invariably carried a gun on board to procure specimens. Large bodies of Gulls and Gannets were round the nets of the boats, on the 9th of November, a few miles outside the Corton floating light; no Skuas were, however, seen. On Monday the 11th a gale set in with heavy squalls from the north-north-east, and continued from different points, with a short intermission on Friday morning, during the whole of the week. Though Gulls and Stormy Petrels were driven helpless on the beach and inland, I recognized but one Skua, an immature Pomatorhine, which was making its way along the shore on the 16th; the bird seemed in an exceedingly weak and feeble condition. On the 20th I steamed out to sea and round the Cross Sands, but met with no Skuas, neither did I learn of any being seen by the fishermen after this date.

In 1879 the weather was excessively stormy in October, and the Skuas suffered in consequence, hundreds and thousands being blown on to our northern and eastern coasts and also inland; numbers were also observed in the Channel. When the first-comers made their appearance early in October, I was not on the coast and did not reach Yarmouth till the 24th. On the 25th, shortly after leaving the harbour, I noticed three adult Pomatorhines, black, with long tail-feathers, flying round the North-Sand buoy and along the beach. As they continued either over the sands or too close in shore for the draught of the steamer, the small boat was launched and I started in pursuit. The birds, however, made their way across the harbour-mouth, and were lost sight of in the mist; though following in the line they had taken as far as Lowestoft, no further signs of them could be obtained. On the 27th several Skuas were met with in the gatways and outside the sands; the birds appeared to be in a weak state, and were in the stage of plumage which probably precedes the adult dress. One fine white-breasted bird with long tail-feathers was floating almost helpless, a mile outside the 'Newarp' light-ship. The weather was dirty and threatening on the 28th, and several Skuas were observed from the deck of the steamboat in the "Would." Heavy squalls of rain and mist, with a strong southerly breeze, on the 30th; a few Pomatorhines were drifting north before the force of the wind, either over the breakers or close along the shore. Here and there one would be seen occasionally settling on the sand-banks, evidently desirous of obtaining rest, though the repeated attacks of the swarms of Grey Crows collected on the beach forced these weary travellers to take wing almost as soon as they alighted. A perfectly black bird with long tail-feathers attracted my attention on several occasions when driven up from the water's edge. Each time the Crows approached with harsh screams and croaks the stranger rose on wing and made his way slowly to windward, returning again after a short interval, drifting in circles before the squalls. Having watched the actions of the whole group for some time, I procured this specimen, as the black form of the adult is evidently uncommon. Though several in various intermediate stages were seen along the shore, I did not recognize another adult of this colour, with the exception of the three observed in company near the harbour-mouth on the 25th. Immature birds of this species were flying north on the 3rd and again on the 4th of November; several in adult plumage were also met with near the 'Cockle' light-ship, floating helpless on the water. One fine white-breasted bird was discerned almost in the breakers; and as it was possible to approach close to the shore at this point (just north of Caister), the steamboat was backed in the endeavour to secure it. The poor bird was alive, though evidently no longer able to contend against the force of the elements, sitting with its head drawn close between its shoulders and washing helplessly at the mercy of the waves. The deidle was over the stern, and another yard would have enabled the capture to be effected, when a heavy sea rolled past and the bird was lost sight of in the breakers. Owing to the surf, it was impossible to launch the small boat and make a search along the shore, where the body would inevitably be cast up in the course of a few minutes. By the end of the first week in November the greater part of the adults had passed south; immense numbers must, however, have perished from the effects of the continued gales. On the 9th, leaving the harbour early, we were at daybreak among the luggers hauling about thirty miles off the land. Not a single adult was recognized, though birds of the year were in considerable numbers, as many as four or five being noticed flying round and chasing the Gulls congregated at the nets of a single boat. As they hovered slowly round while awaiting their prey, it was easy to distinguish several different shades in the colouring of the plumage; and I succeeded in procuring as many specimens as I needed of both the dark and light varieties. Steaming slowly towards the land after midday, I noticed clouds of Gulls following in the wake of some of the boats which were also making for the harbour. In a few instances where the operation of cleaning the nets was being carried on, the Kittiwakes continued in attendance as far as the "roads," flying round in circles and swooping down to the water as soon as the smallest particle of fish dropped overboard. A short distance outside the busy throng of Gulls, small parties of two or three Skuas kept steadily in view, though the moment any particularly inviting morsel had been seized they darted off at once in pursuit of the captor. On the 13th the wind blew strong from north-north-west, and two or three immature Pomatorhines were flying before the gale along the sands. On the 18th out again among the luggers while the nets were being hauled. The Skuas appeared to have mostly passed on, but few in comparison with those lately observed being met with. Young only were noticed; by feeding the Kittiwakes with liver and fish hundreds were drawn in attendance on the steamboat for over twenty miles. The sight of this dense gathering perpetually dipping down for food induced every Skua that hove in view to bear down at once on the swarm of birds. In one or two instances I remarked that the most ravenous of the Skuas swooped down close to the stern and seized for themselves the food that was flung over. Hearing that the sprat-boats above Lowestoft were getting immense hauls of fish, we steamed on the 28th inside the sands as far as Southwold; Kittiwakes in thousands were round the boats and also a short distance out to sea. Skuas were still plentiful and numbers collected when the Gulls were supplied with food, young only being noticed. They proved especially tame, many hovering just above the deck while the fish were being thrown over; the hungry birds afforded such an excellent chance of examining their stages of plumage that I was saved the necessity of committing any useless slaughter. The following day, the 29th, our course was again along the coast, and much the same birds were encountered; this was the last occasion I was at sea off the east coast during the season of 1879.

While travelling by rail, on the 1st of December, across the Suffolk marshes, an immature Skua was passed, sitting on the snow within twenty yards of the line. The bird was apparently worn out, and had probably been carried inland, being utterly incapable of contending against the strong easterly breeze.

On reaching the south coast, I learned from the fishermen that earlier in the season many Skuas had been met with in the Channel; they had, however, by this time all apparently passed on towards the west. Though I was almost daily at sea during December, not a single specimen was observed.

Since the autumn of 1879 I have not joined the herring-fleet in the North Sea, in order to obtain a further insight into the movements of this species; according to the information gathered from the best authorities among the crews of the fishing-vessels, the numbers that approached the land during the last two or three seasons have been exceedingly small.

During a terrific gale of wind from the south on the 24th of October, 1882, a small party of Skuas of this species were blown inland at Shoreham and settled for a time on a stubble-field; though I did not succeed in obtaining a close view, it was obvious that the whole were in immature plumage. It is seldom I have met with the Pomatorhine Skua at any distance from salt water; a couple of immature birds, however, were observed tearing at the repulsive scraps of decaying meat on a heap of town-refuse carted out from North Berwick to one of the neighbouring farms in the autumn of 1863. These youngsters were busily engaged in company with some juvenile Herring- and Lesser Black-backed Gulls; the whole party appeared most amicable, not the slightest attempt at robbery being attempted while I watched their proceedings.

I am of opinion that Pomatorhine Skuas attain the age of five years before the perfect adult plumage is assumed. Judging from observations on the various stages through which the birds I kept in confinement passed, it appears evident that these Skuas are seldom seen off our coasts while exhibiting the plumage of the second or third years. But a single specimen in the dress of the third year has come under my notice, this rarity having been procured by the master of a Yarmouth fishing-lugger during the autumn of 1872. Meeting the skipper on the quay shortly after landing, I learned he had one or two strange birds, killed during his last voyage, that he would like me to examine. I was well aware he had been afloat for at least a fortnight, and consequently imagined they might be getting unpleasant; being assured, however, that they were as fresh as when first shot, it was settled he should bring them down for inspection in the evening. The specimens, which proved to be an immature Gannet, a Pomatorhine Skua, and a bird whose identity I could not satisfactorily determine, stunk in such a horrible manner that I was forced to request him to take them away and allow me to see them by daylight, when an opportunity might be obtained to examine the stranger more closely out of doors. On arriving at the man's house on the following morning, I discovered that during his absence from home they had been sold by his wife for sixpence to a musician, as an old and two young Harnsers *, and were at that moment being cooked for the Sunday's repast, much to the disgust of his neighbours in the row, who were almost poisoned by the stink. That the unknown bird was a Skua there was little doubt, though the age, and even the species, I was unable to decide at the time. Since keeping these Skuas in confinement and noting their changes, I have no doubt that the bird was a light-coloured specimen of the Pomatorhine in the third year's plumage. As far as I have been able to ascertain, this stage has neither been figured nor described.

The distinction between the dark and light forms of this Skua is apparent in the earliest stages; I noticed a great difference in the shades exhibited by several specimens obtained in the first plumage on the 9th of November, 1879. Though birds showing white breasts when adult are by far the most numerous, it is easy to trace the two forms in every stage of plumage through which they pass.

In November 1879 I procured at Yarmouth four Pomatorhine Skuas alive (three immature and one adult), all being captured at sea when reduced to the last extremity by the protracted buffetings of the gales. The mature bird only survived confinement till the following April; I was, however, enabled to ascertain that the adults of this species undergo a change in winter, even after the full plumage has been assumed. This specimen was perfectly black, the plumage being darker and more glossy than on the black form of the adult figured on Plate III. Round both sides of the neck several feathers edged with a yellow or gold tint showed up most conspicuously; similar markings were also noticed on two or three other black Skuas which I examined when fresh caught, having been picked up in the North Sea by the crews of the fishing-luggers. At the time of its death the black adult was entirely altered in appearance; the throat, breast, and underparts were almost white, and light grey feathers were showing thickly round the neck; the back exhibited also a few fresh

* Norfolk name for the Heron.

feathers edged with white. When first taken but one of the elongated feathers remained in the tail, and this was lost shortly after. The three others captured at the same time were, as previously mentioned, birds of the year; one escaped, owing to its wing not having been clipped, but the other two remained perfectly healthy in confinement till the 19th of September, 1882, when both were unfortunately drowned. From their earliest stages one of these captives was considerably darker in colouring, and this distinction was retained through every change till the end. When once a suitable groundwork for their enclosure had been obtained (turf grown on a foundation of chalk) they were not troubled with corns or swellings on their feet, and the old wounds from which they suffered at first speedily disappeared. The cries and screams they uttered were most amusing, and the names of Punch and Judy were bestowed on account of their extraordinary vocal performances. Any wandering proprietor of the old familiar show of Punch would have been wild with jealousy at the "Toby, Toby, Toby," ending with a prolonged squeal and a whistle, which, when excited, one or the other of these strange birds would occasionally give vent to. The captives thrived well on herrings, mackerel, or sprats, their actions while feeding being exceedingly singular. If one happened to seize a portion of food too large to be swallowed with ease, he would call loudly, when his companion at once ran rapidly up, and clutching hold of one end of the fish, each would tug lustily till the whole was divided, when the parts were consumed by the pair in the most amicable manner. This curious performance was now and then repeated half a dozen times during the same meal.

Though Pomatorhine Skuas are considerably more powerful than the Arctic, I believe it is seldom they pursue any of the larger Gulls when in quest of prey, Kittiwakes and the various species of Terns being most frequently attacked by both the Arctic and Pomatorhine. On one occasion during the autumn of 1871, I watched an adult flying in such a manner as to give the impression that it was in chase of a Heron along the shore near Canty Bay; the awkward long-legged wader was evidently terribly scared, shrieking and vainly attempting to avoid the swoop of the dashing sea-bird. As far as I was able to ascertain, there was little cause for alarm, as the Skua passed rapidly on, apparently only endeavouring to hustle the Heron out of his course.

As "black" or "dirty black Allans," Skuas are spoken of by most of the Scotch fishermen; occasionally I heard this name made use of by a few of the men sailing in the east-coast luggers out of Yarmouth and Lowestoft, though to the majority they are known as "Molberries"*. Along the Cornish coast they are usually called "Tom Harries," while the seafaring fraternity belonging to the Sussex ports bestow other titles on these robbers; the names they give, however, being derived from entirely mistaken ideas concerning the habits of the birds, are not worth recording.

While in correspondence with the light-ships off the east coast in 1872 and the following year, I ascertained that Gulls occasionally came in contact with the lamps, though it was by no means a common occurrence for any species, with the exception of Skuas, to be taken in this manner. The mate of the 'Newarp' informed me that he had once found as many as three Skuas on deck during his watch—one of which, a large brown-coloured bird, probably a Great Skua, mistaken in the dark for a fowl while lying disabled in a corner, inflicted a most severe bite on his hand. From all I could learn, the species usually taken were either Pomatorhine or Arctic, in the various immature stages of plumage.

During the autumnal migration I repeatedly remarked that the order in which these Skuas passed southward was as follows:—birds in the intermediate stages appeared first, followed a few weeks later by the full-plumaged adults, while the juveniles invariably brought up the rear. Though a few adults, whose presence has been duly recorded in my notes, pass early in the autumn, they do not form a tenth part of the individuals observed off our coasts at that season.

* I was utterly unable to discover any derivation for this strange name.

A few words descriptive of the Plates may perhaps be of service:—

In Plate I. are figured two young birds of the year, representing the light and dark forms, shot at sea off Yarmouth on the 9th of November, 1879.

The state of plumage shown by the light form of this Skua during the third year is given in figure 1, Plate II., the sketch having been taken from one of the captives towards the close of the summer of 1881. This specimen corresponds precisely with the Skua previously referred to as shot by the master of a fishing-lugger off the Yorkshire coast in October 1872. During all the years I have knocked about in the North Sea not a single Skua in a wild state exhibiting this stage of plumage has come under my observation; it is probable that the young birds, after reaching their winter-haunts (wherever they may be), seldom visit our shores during the two following years. Figures 2 and 3 in the same Plate are also taken from the birds kept in confinement, and show the state of plumage at the time of their death (10th of September, 1882), when a few months over three years of age. Pomatorhine Skuas in imperfect plumage, as previously stated, are usually met with off the Scotch coast somewhat in advance of the mature birds, many that are seen in August bearing a strong resemblance to those represented by figures 2 and 3.

The dark and light forms of the adults are shown in Plate III., the former having been obtained along the shore near Yarmouth on October 30th, 1879, and the latter at sea off Caister the following week.

ARCTIC SKUA.

STERCORARIUS PARASITICUS.

This species, also known as Richardson's Skua and Black-toed Gull, is a summer visitor to many parts of the north of Scotland and the adjacent islands. During spring and autumn these birds have come under my observation on several occasions a few miles at sea in the English Channel, as well as at times in great numbers off the south-east coast of Scotland and also in the North Sea.

Of late years there has been a considerable falling off in the numbers of these Skuas resorting to many of the localities formerly frequented in the Highlands. In some parts the birds, if not totally exterminated by keepers, have been greatly thinned down; while in others so constantly have their nests and young been pillaged by collectors and dealers, that it is a wonder any survivors remain.

This species passes north towards its breeding-quarters probably in April. In 1875 I was out frequently in the Channel from 6 to 12 miles off the Sussex coast, and on several occasions between the 11th and 23rd fell in with single birds as well as small parties; they were especially numerous on April 21st and 22nd, wind on both days light from the east. A few obtained as specimens were in full breeding-plumage with long tails; others, apparently in the same stage, were minus the long tail-feathers; not a single bird was seen in the immature stage.

By the end of the first or second week in May I have met with these birds on the moors in Caithness. In many instances in this locality I remarked that their favourite resorts appeared to be either on or in the neighbourhood of the "floes" *. On one of these spots, which extended nearly two miles in length by almost the same in breadth, hundreds of Lesser Black-backed Gulls were nesting. Though a few Skuas had reached this part by the middle of May, and evidently taken up their quarters, I could find no signs that nesting had as yet commenced. The birds were usually at no great distance, and, after wandering for a time among the pools on the floe, they were sure to be detected making straight for the intruders on their domain. After sweeping round for a few minutes, and giving vent to their anger in harsh screams and snaps, they would, as a rule, gradually draw off; their attack was by no means so decided and vindictive as when they had thoroughly entered upon their breeding-operations. The earliest date on which eggs were noticed in this locality was on May 31st, the nest being placed on the open moor a short distance from the floe. The pair of birds in this instance were remarkably shy, the female being especially careful, and seldom venturing within eighty or one hundred yards; as they, however, resolutely refused to quit that part of the moor, it was evident their treasures were concealed near at hand. The nest, which was merely a slight depression in the moss, with a few strands of dried grass by way of a lining, was not discovered till we retired some distance; and having posted a couple of keepers four or five hundred yards apart, I took another station myself which

* The "floes" are large boggy tracts of moor, full of small pools of deep black peaty water. The larger pools usually contain diminutive islets; and the bottom being too soft to admit of wading, they are safe spots for birds to nest on. The india-rubber boat was most serviceable in this locality.

commanded the whole of the floe. After carefully watching the male through the glasses for nearly an hour, he was at length marked down. When driven up from the eggs he showed far greater animosity, dashing down within the distance of ten or twenty yards, but not venturing to make the slightest attack on us or on a retriever following close to my heels. Being anxious to procure both male and female in order to investigate the colouring of the sexes, I experienced but little difficulty with the former; but after the death of her mate the female became still more wary, and it was not till after several hours' delay that a shot was obtained: these birds are figured in Plate III. A week or so later other nests were met with in the same district: in no instance were they placed in close proximity to one another or, indeed, within some hundreds of yards of the quarters of the Gulls. The birds varied considerably in their attitude towards intruders, some few pairs being far more demonstrative than others, though I failed to recognize the audacity occasionally ascribed to them by some writers.

The young in the earliest stage were covered with a soot-coloured down. After the eggs were hatched the parent birds manifested increased disapproval at the invasion of their haunts.

In most of the pairs which came under observation, one of the birds was in the dark- and the other in the white-breasted plumage. A few pairs were noticed in which both birds exhibited the light-coloured plumage, though I was unable to detect a single couple where both were black.

In this district the principal food of the Arctic Skua is without doubt the young of the salmon; no other fish were detected in any of the specimens obtained on these moors. The stomach of every bird that was inspected contained from two to five smolts.

The Common Gulls, which breed in numbers round some of the lochs in the neighbourhood, capture these silvery little fish in the shallows of the rivers, and are forced to deliver up their prey when attacked.

On two or three occasions, while following the course of a gully or burn across the moors, I came suddenly, within the distance of one hundred yards, upon a few pairs of Skuas resting quietly on the moss-covered mounds so commonly seen on these flats. On examining the spots from which they rose, I discovered several castings composed entirely of small pieces of egg-shell. In almost every instance the fragments, so far as I was able to judge, appeared to have belonged to the eggs of sea-fowl, the Lesser Black-backed or Common Gull (in all probability the latter) having been the principal sufferers. On a subsequent occasion a more careful examination of the remains led to the belief that the nests of Grouse and Plover had been plundered, though it was by no means easy to form a decided opinion. These mounds were evidently a favourite resort to which the robbers retired for repose in order to digest their food. When viewed at a distance through the glasses, I now and then detected some of their number squatted on the moss with their heads either drawn down between their shoulders or turned over and partially concealed among the feathers of the back: one or two were, however, invariably on the alert, and the scream of a passing Gull, or even the low whistle of a startled Plover or Dunlin, would instantly bring the whole party to their feet.

During the latter part of May 1868 a few single Skuas of this species were occasionally seen sweeping over Loch Maree and other large sheets of fresh water in the west of Ross-shire. I did not, however, meet with any breeding-quarters in that locality, and it is probable that these stragglers were from the Outer Hebrides. Though the tints on the breasts of the birds observed in this district varied from white to a pale clouded grey, I did not recognize a single dark specimen.

Early in August a few of the Skuas in perfect breeding-plumage usually show themselves off the south-east coast of Scotland, gradually increasing in numbers as the autumn advances. My notes jotted down during several years contain references to the appearance of these birds on the coast of East Lothian at about the same date; the numbers, however, that approach the land vary considerably. In 1874 I paid particular attention to this species during a three months' residence at Canty Bay; a few extracts from my journal may give some idea of the habits of Skuas while on their passage south for the winter.

The earliest entry referring to these birds is under date of August 11th to 13th. Wind during these days exceedingly squally and from all quarters. Two or three small parties, all apparently Arctic Skuas, flying slowly head to wind, were noticed halfway between the Bass and the land. On the 19th, fine with light west wind, several Skuas passed the boat a few miles outside the Bass; the distance, however, was almost too great to judge accurately as to the species, although I was able to detect some Pomatorhine. A couple of days later, with the wind east, Skuas of both species were busy round the boat while shooting long-lines. The 22nd was stormy, with a strong breeze from the west; starting early from Canty Bay, we sailed alongshore halfway to Dunbar; but though thousands of Gulls were resting along the rocks, only two Arctic Skuas were seen. In the afternoon, while hauling the long-line, which a few heavy congers had succeeded in entangling among the rocks in such inextricable confusion that it had to be cut four times and grappled ere the whole was recovered, numbers of Gulls gathered round the boat, and in pursuit of these came several Arctic Skuas, most of them being dark with long tails. Dead Kittiwakes, I soon learned, were exceedingly useful as decoys; a few flung up in the air would usually draw any passing Skuas; it was then possible to examine their plumage, and with but little trouble any desirable specimen could be procured. While fishing alongshore on the 25th, Skuas, principally Arctic, were passing east and west in pursuit of Gulls. For several days, commencing on the 26th, we joined the fleet of boats that sail from North Berwick, shooting their long-lines either between Fidra and the Bass or other parts of the Firth, according to the wind. The lines are hauled shortly after daybreak; and as soon as the operation commences, Kittiwakes collect in hundreds round the boats. The smaller haddies and whiting, being of no value for the market, are usually dashed by the fishermen against the side of the boat in order to free them from the hook, and in this manner save trouble previous to coiling the line. The hungry Gulls, evidently well acquainted with the whole of the proceeding, meanwhile hover round within a few yards, and whenever a diminutive fish emerges from the water, at once draw nearer, and darting down secure their prey before it reaches the waves, snatching many within a foot of the boats. A short distance outside this busy throng the Skuas are eagerly awaiting their share. No sooner does a Kittiwake, after having secured a fish or two, attempt to make off, than it is assailed and forced to disgorge its meal. So ravenous are these pirates that two or three may often be seen in pursuit of one unfortunate Gull. I remarked that Skuas would not commence an attack on the victim they had singled out while he was on the water. A Kittiwake that had evidently fared sumptuously, and was either too full or too frightened to rise, attracted my attention by his endeavours to avoid three Arctic Skuas swimming round and frustrating every attempt at escape. Whichever way the poor bird turned he was invariably headed, though his persecutors neither molested him in the slightest degree nor even approached within the distance of two or three feet. How this evidently well-planned conspiracy would have terminated I had no opportunity of learning, as a number of fresh Skuas arriving round one of the other boats, I was forced to take an unfair advantage and turn over these birds by a sitting shot, a couple of them being in a state of plumage I was anxious to examine. Though the first-comers that made their appearance along the coast were all in adult breeding-dress, there were by this time several exhibiting a mixed state of plumage; no young of the year had, however, as yet put in an appearance. But few Skuas showed themselves round the boats on the 28th; two or three Pomatorhines were, however, shot, as well as a couple of especially black-plumaged Arctic. These two birds, I ascertained, were male and female; and it certainly appears that there is no rule as to the colouring of the sexes. On the 29th the smaller Skuas were again numerous, though Pomatorhine were noticed at a distance, nearly all flying west at a considerable height. One very small white-breasted Arctic Skua, a male, was shot; the plumage was perfect, but the long tail-feathers were missing. Another well-marked specimen, a white-breasted bird with long tail-feathers, proved to be a female. Constant squalls of rain and wind from the 31st to September 5th, several Skuas of both species being seen alongshore. The weather was fine and still on the 8th, and soon after daybreak we were again out with the boats hauling long-lines; Arctic Skuas appeared in great numbers, though not above half a dozen Pomatorhine were observed. The Firth from Fidra to the Bass Rock was swarming with Skuas, and I repeatedly noticed two or three pursuing one unfortunate Kittiwake; they often continued their attacks for a long distance, driving the bird at times high into the air. I subsequently remarked that the Skuas so persistently following the victims they had singled out were in immature plumage: probably they were scarcely up to their business through want of experience, and possibly even tackled birds having no food to disgorge. A Kittiwake does not require to be hard-pressed in order to eject any recently taken prey, and the screams the affrighted creatures gave vent to while assailed and buffeted could never have been uttered by throats distended with food *. A squally morning on the 9th was succeeded by a fine afternoon; while about a quarter of a mile south-west of the Bass, late in the evening, a flock of between forty and fifty Skuas passed the boat, flapping slowly towards the west. The party were flying close to the water and afforded an excellent chance for careful inspection. With but a single exception, a small dark bird which proved to be an immature Buffon's Skua, the whole number appeared adult Arctic, all, with two or three exceptions, still retaining the long tail-feathers, the dark and light varieties being represented in about equal numbers. So thickly were they making their way that the charge of shot which stopped the juvenile in their ranks (the only specimen I was anxious to examine) brought down as well a couple of white-breasted and one dark adult in perfect plumage. The presence of this small stranger in a flock composed entirely of Arctic Skuas, unaccompanied by any of their own young, appears somewhat singular. After several exceedingly squally days, during which the greater part of the Skuas passing within sight of the shore were too distant to be identified, we again got to sea on the 16th and found Skuas numerous up the Firth. Between the islands of Fidra and Ebris I obtained a couple showing a curious mottled and barred state of plumage, evidently the intermediate stages of the black and white varieties. From the observations I was enabled to make concerning their various stages of plumage while keeping Pomatorhine Skuas in confinement, it is probable that these birds were both in the third year. It is, however, possible that the adults of this species also undergo certain changes after the breeding-season—in which case it would be hard to say whether they were immature or adults in winter plumage. A fresh breeze from the north on the 17th; the boats were hauling lines between Fidra and the Lamb and, as usual, several Skuas were seen; one bird only was obtained, a particularly dark specimen with long tail-feathers, in excellent condition. While on our way back, halfway between Craig Leigh and the Lamb, we encountered a large school of whales numbering at least a couple of hundred; several passed close to the boat, rolling and diving, and apparently chasing one another both above and below the surface. I was on the point of trying a shot with a heavy gun at the head of one of them, and the younger portion of the crew gathered round in the greatest excitement to watch the effect. The two old hands were seated aft, apparently unconcerned, when Kelly, who was steering, cast a glance at the waves as he shook from his arm the spray that had broken over the quarter, and quietly remarked, "The watter's gay an' cauld the morn." Now Kelly and MacLean were two aged and weatherbeaten salts, who had sailed for years in the Arctic whaling-fleet, and survived many a tussle with the monsters of the deep in Baffin's Bay and Davis Straits. I doubted not they understood the manners and customs of this unruly family, and that such an insult as a charge of shot would probably result in our craft being smashed and the crew dispersed floundering in the water. As the tide was rushing out of the Firth like a mill-race, and there was not a sail within three or four miles, I soon came to the conclusion that discretion was the wisest plan, and allowed the monster to roll on his way unmolested. Two or three others, however, immediately after dashed past us so closely (plunging, in fact, in a playful manner right under the boat), that I began to fancy we might be favoured with a passing whisk of their tails without having intentionally offered the slightest insult. From the 18th Skuas were seen daily, the Arctic being by far the most numerous; Pomatorhine usually observed flying high in the air, either due east or west. On the 28th, with a fresh southerly

* Having kept various kinds of Gulls in confinement for years, I have had constant opportunities of studying their habits: when terrified or even suddenly aroused after feeding, they endeavour to cast up the contents of their stomachs.

breeze, sailed up the Firth as far as Aberlady Bay; several Skuas met with, but only a couple of specimens procured, both immature Arctic birds of the year. The plumage exhibited a warm brown tinge mottled with black. The legs and feet were a dull black marked with a pale bluish-white line down the tarsi from below the knee, and extending about a quarter down the webs. Several Skuas seen daily along the coast till the end of the month. During a terrible gale of wind from the south-west, which continued for several days from the 2nd of October, immense numbers of Gulls and Skuas were visible in the bay. On the 7th and 8th the weather moderated, and we succeeded in getting out to sea; a few miles off the land, from Fidra to halfway between the Bass and the Isle of May, Skuas were plentiful, following in the wake of the large flocks of Kittiwakes. Immature Arctic Skuas were now arriving in the Firth, the first-comers having been mainly composed of adults.

According to my own observations, Arctic Skuas reach our north-east coast in the autumn, considerably in advance of Pomatorhine, and for the most part move southward at an earlier date. When out in the North Sea off Yarmouth, in company with the herring-fleet, throughout the latter part of October and November 1872, and again in 1879, Pomatorhine Skuas were constantly seen, scores or even hundreds being occasionally fallen in with in the course of a day, though not more than a dozen Arctic were recognized during the whole of either season.

The Arctic Skua, according to my experience, is seldom, if ever, seen in pursuit of sea-fowls larger than Common Gulls, Kittiwakes, or Terns. But few instances where remarked where the former species has been molested at sea, the unfortunate Kittiwake being far more heavily taxed in order to contribute to the support of these rapacious thieves. While in their summer quarters on the moors they not unfrequently harass their neighbours as well as any feathered stranger approaching their haunts; under such circumstances their noisy attacks are not commenced with a view of procuring food.

Judging from the various stages of plumage exhibited by this species, it is probable that the perfect adult dress is not assumed before the fifth or possibly even the sixth year. Never having kept these Skuas in confinement, I do not feel justified in offering any decided opinion on the subject; it must, however, be allowed that there is no rule as to the colouring of the sexes. The dark and light varieties were without difficulty to be traced in many specimens I examined during the immature stages, the tints in even the first feathers of the young indicating, in some instances at least, their future colour. The distinction between the dark and light varieties in the earlier stages of this species is by no means so conspicuous as in the Pomatorhine Skua. In the same manner as their larger relatives, numbers of these birds in adult dress are seen during autumn without the long tail-feathers. I also remarked this fact in a few specimens exhibiting mature plumage procured in the Channel when on their way to the north in the spring. I cannot call to mind a single instance of observing birds on the moors in the north during the summer without these feathers.

As to the winter dress of the adult Arctic Skua, there is doubtless much to be learned. It is, I consider, quite possible that specimens met with late in autumn, and considered immature, may be adults undergoing the change to winter plumage. It would be interesting to ascertain the length of time that the light blue marking on the tarsi and webs is retained; it would then be more easy to form an opinion.

The majority of the fishing-population with whom I have conversed appear to recognize no distinction between the various species of Skuas frequenting our coasts, the whole family being apparently known by the titles mentioned under the heading of the Pomatorhine. It is possible the birds may occasionally be referred to as large or small, but no further difference is recognized. One or two masters of smacks or fishing-luggers (old gunners who still take an interest in wildfowl and all feathered strangers) have at times, however, rendered me great assistance by stating the fishing-grounds on which birds in some peculiar stage of plumage were to be found. Acting on their information, I have more than once steamed twenty or thirty miles to some distant bank in the North Sea; and within a few miles of their reckoning, the objects of our search were invariably met with.

The following are the stages of this species figured in the Plates, together with the dates on which they were obtained:—

Plate I. A young bird of the year, shot September 28th, 1874, in Aberlady Bay in the Firth of Forth.

Plate II. Intermediate stages of dark and light form (probably in the third or fourth year, judging from the changes exhibited by Pomatorhine Skuas kept in confinement), shot September 10th, 1874, between the islands of Fidra and Ebris in the Firth of Forth.

Plate III. Adult male (light form) and female (dark form), obtained at their nesting-quarters in Caithness, May 31st, 1869.

Plate IV. Two adult females, shot in the Firth of Forth on August 28th and 29th, 1874.

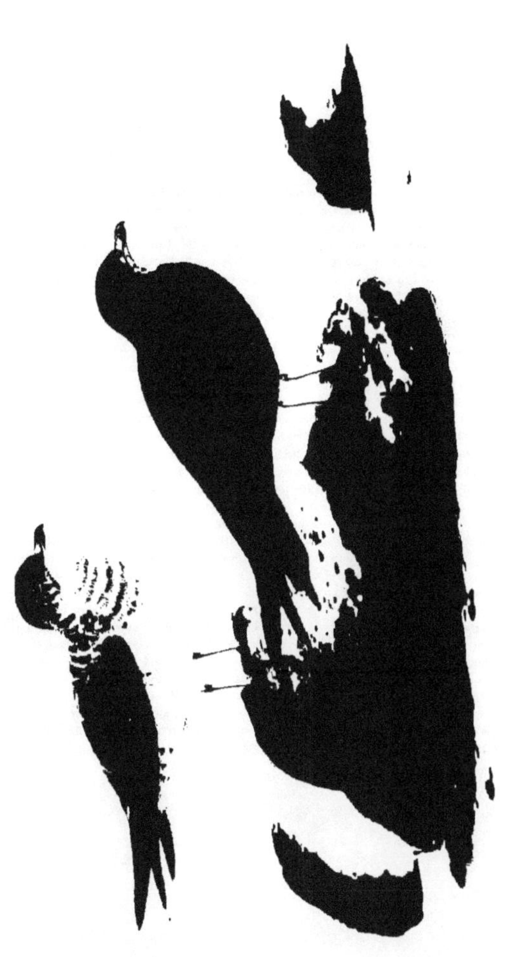

LONG-TAILED SKUA.

LESTRIS LONGICAUDATA.

If this species formerly nested in the more remote districts of the Northern Highlands, it is probable that all the breeding-haunts are now deserted; I have visited a few of the localities which these birds are said to have frequented in days gone by without meeting with a single specimen. The Long-tailed Skuas that pass along our coasts during summer or autumn while on the passage to or from their nesting-quarters in the far north are seldom, if ever, seen in such numbers as the Arctic or Pomatorhine; on no occasion have parties consisting of more than two or three flying in company come under my observation. It is probable, after the manner of the Arctic Skua, that these robbers levy a tax on the smaller Gulls when needing food, though I am only enabled to speak with certainty as to their attacks on Terns, but few opportunities of studying their habits having fallen to my lot.

Early in June 1865, while brought up fishing to the north of the Bass, I watched for over an hour a pair of these birds exhibiting the most perfect plumage; after flying round in the vicinity of the rock for a time, they settled down near a broken spar floating past on the tide, one at length rising up and alighting on the timber. Though resting quietly and apparently unconcerned, these handsome strangers were evidently on the alert, and exceedingly wary when approached: having attached a buoy to the cable, an attempt was made to pull within range; both birds, however, instantly rose on wing and proceeded up the firth towards Craig Leith, returning shortly and pitching down near the same spot. On the first signs of danger, the actions of the Skua perched on the wreckage were most singular, and I much regretted that there was not an artist at hand to take its portrait from life; starting up at once from repose, it indulged in a comfortable stretch, a position eminently suited to show off its striking plumage and elegant form to the fullest advantage. Taking a couple of steps, the bird bent its neck downwards, the head being at the same time stretched forward, the wings were next raised and opened and the feathers of the tail elevated, the whole performance being carried out in a slow and deliberate manner; then turning and uttering a harsh cry, it followed rapidly in pursuit of its mate already on wing. Judging from the behaviour of this pair, there could be but little doubt that they were male and female on the way to their breeding-grounds in the far north, their actions on the water being similar to those of many sea-fowl watched at this season—an immense amount of swimming round, bowing, and chattering being indulged in. A second attempt to obtain a shot proved also a failure; and again bringing up at our old quarters, I waited on the chance that the birds, who continued on wing for the rest of the time they remained in view, might work within range of their own accord. Kittiwakes in thousands hovered over the tideway in the bright sunshine, occasionally making their way to and from the ledges in the face of the cliffs, while clouds of Gannets and straggling parties of Guillemots were circling round or heading straight out for the open sea in quest of prey. Through all these busy swarms the pair of Skuas held a steady course, turning neither to the right nor to the left; at times they appeared to wake up—dashing down with a rapid swoop towards the water they skimmed slowly over the surface, then rising again, repeated the performance after a short interval. It was evident that the Kittiwakes did not appreciate their society, making rapidly off when approached, though no attack was offered; at length some ten or a dozen Terns flapped quietly down the firth, when, with a sharp and piercing cry, the pirates were off in pursuit; I failed, however, to ascertain the result, owing to the countless numbers of sea-fowl sweeping around. The two Skuas were shortly after seen again near the rock hunting over the same course, being eventually lost sight of while working along the coast towards the east.

After continued rough weather for several days in November 1870, I received word that a small party of Skuas were sheltering on the water below the West Pier at Brighton; on reaching the spot and putting off in a boat, I discovered that the majority had already taken their departure, but one bird (which proved exceedingly unsuspicious of danger) remaining. Its plumage having been carefully examined through the glasses, the specimen was procured without the slightest difficulty, the bird being evidently rendered almost helpless by the protracted gales. It is by no means an uncommon occurrence for Skuas to resort to the smooth water in the vicinity of the pier during storms in autumn; as many as ten or a dozen may occasionally be seen, at times much distressed by the severity of the weather, but rapidly pulling round as the wind drops.

On the 17th of May, 1872 (a strong breeze from the north-east with squalls of rain), I noticed early in the morning a pair of immature birds, in very dark plumage, flying round the boats brought up off the fish-market at Brighton; the pair appeared perfectly fearless, but succeeded in keeping out of range. Towards evening the birds (or a similar pair) were met with a short distance at sea off Shoreham and again escaped. I remarked that these Skuas hovered over the refuse floating along the shore and round the fishing-craft moored off the town, evidently searching for food themselves, no attention being paid to the movements of several Common Gulls engaged in a similar manner.

A young bird of the year was shot near the Bass Rock on the 9th of September, 1874. This specimen was in company with forty or fifty adult Arctic Skuas; the whole party flapping close past the boat, ample opportunities were afforded for examining them satisfactorily. Though a sharp look-out was kept during the succeeding week, not another bird of this species was recognized either in the firth or for several miles at sea. I am inclined to believe, from the numbers of specimens brought in by fishermen from the distant banks in the North Sea, that the usual course of this species, while on their way south at the time of the autumnal migration, lies considerably to the east of the British Islands.

But little change appears to be exhibited by young birds in the spring of their second year. In May 1875, while at sea in the Channel a few miles off Shoreham, a pair in immature plumage hovered for some time round the boat; and both, I was perfectly convinced, corresponded precisely with young obtained in their first autumn. These interesting juveniles proved exceedingly tame; on throwing out small portions of fish-liver, they dipped down to the surface of the water and seized the pieces within the distance of three or four yards. The slate-blue markings on their feet and legs being clearly seen and noted down while they swept round the boat, rendered it unnecessary to obtain a single specimen. So confiding did they eventually become after remaining for an hour or two in our company, that on offering a supply of fish-liver on the blade of an oar, both birds paddled up and seizing the pieces one by one, swallowed them without exhibiting the slightest signs of alarm.

The autumn gales of 1879 forced thousands of Skuas out of their usual course, immense numbers being blown ashore in a disabled condition along the east coast. Several of this species were procured by the fishermen and shore-shooters, in both adult and immature plumage; two specimens, however, were all that came under my observation. During the last week in October I noticed a young bird of the year passing along the shore near the harbour-mouth at Yarmouth, and a few days later another of the same age was met with at sea a short distance off Caister. This poor traveller was so worn out by the protracted buffetings of the late storms that I was in hopes of effecting a capture in order to rear it in confinement and note the changes of plumage. When first seen it was on wing, skimming before a fresh northerly breeze in a line with the coast; at length it turned head to wind and, after circling round for a few moments, settled on the water; then drifting with the tide through one of the "swashways"*, we were enabled to steam ahead and await its approach. The heavy swell rendering the launching of a boat a work of time, we were forced to make use of the "deidle," when, just as the bird was carried within reach, a breaking sea cast it under the paddle-wheels, and a few moments later its lifeless body was discerned from the bridge washing over the sands.

The immature bird figured in the Plate was shot on September 9, 1874, near the Bass Rock. The specimen exhibiting a mixed plumage, after having been almost driven ashore under the West Pier at Brighton, was procured a short distance at sea in November 1870; it appears doubtful whether this bird shows one of the intermediate stages, or is an adult undergoing the transformation into winter plumage. From observations made respecting the changes of other species of Skuas while in confinement, I am of opinion that the latter is most probably the case.

* Small channels through the sands, not considered navigable, but occasionally made use of by the tug-steamers and fishing-craft.

MANX SHEARWATER.

PUFFINUS ANGLORUM.

It is only on the open sea that I have met with an opportunity of watching the habits and studying the actions of the Shearwater; never having visited their breeding-quarters on any parts of our coasts, the information I am enabled to give concerning this species is exceedingly scanty. While steaming off the Scotch coast in the Firth of Forth, within a few miles of the Bass and the May, as well as in the North Sea some twenty or thirty miles off the land, I frequently observed Shearwaters during autumn. On the Minch and about the entrance to some of the saltwater lochs among the islands I caught a glimpse on two or three occasions of single birds and small parties skimming over the waves in the evening twilight during summer; my time while in that district being, however, fully occupied I was unable to explore the rocky islets to which they were said to resort.

A strong breeze at times brings large numbers of Shearwaters into the Firth of Forth. On the 14th of August, 1874, an attempt was made to reach the May in order to make observations on the various species of sea-fowl resorting to the island; a heavy sea rolling into the Firth with a breeze gradually freshening into a gale from the north-east, however, put an end to all hopes of landing for the remainder of the day at several of the spots we intended visiting. After steaming under the shelter of the south side of the island to the entrance of several of the caves, we turned slowly back towards the south shore of the Firth, passing hundreds of Gannets plunging down for fish, utterly regardless of the swell. Where the Geese were most numerous several Shearwaters were sweeping round with their singular and characteristic flight, hovering at times for a moment and apparently seizing some small particles of food off the waves. Whether they merely dipped into the water while turning, or were engaged in securing prey, I was unable to ascertain satisfactorily, the rolling and pitching of the steamboat and the flying spray rendering it impossible to use glasses to any purpose. A couple of shots were fired and one bird knocked down, though almost immediately lost sight of in the broken water. Parties of Shearwaters were met with repeatedly till within about a couple of miles of the Bass; I have seen them on one or two occasions only between the rock and the shore. When viewed at the distance of two or three hundred yards, the flight of these dusky-looking birds as they skim rapidly over the waves, checking themselves to hover for a moment, and the next darting off and circling round, appeared to bear a strong resemblance to the movements of the humming-bird moth. On bringing up off Canty Bay it was evident that, owing to the heavy sea and the surf beating over the rocks near the landing-place, to make an attempt to reach the shore with either of the boats on board was decidedly risky. One of the crew had just suggested that we should steam up the Firth as far as North Berwick, where, he informed us, "fine intelligent horses and machines"* could be procured to convey us back to Canty Bay, when a large salmon-coble with a strong crew was observed putting out to sea; it was a heavy pull making headway against the swell, but at last they were alongside the vessel. Awaiting an

* The conveyance one hires in this part of East Lothian, whether landau, waggonette, or dog-cart, is invariably termed a "machine."

opportunity as the boat rose on a wave, there was little difficulty in dropping on board, and the landing was eventually effected without shipping a drop of water. The combined fishing and salmon crews quartered in the village having rushed down into the surf, our craft was hauled high and dry without delay, and the anticipated drenching avoided. On the 15th, the weather having moderated, we were again at sea, this time in a fishing-lugger, with hands sufficient to pull four long oars if required, and several Shearwaters were observed a mile or two outside the Bass. The birds proved exceedingly restless, seldom remaining quiet for any length of time, constantly rising, hovering for a few moments over the water, and then dashing off on an extended flight towards the open sea; the wind increasing after midday, we were forced to put back without obtaining a single specimen. A few days later (August 19th) the weather was all that could be desired, a light westerly breeze extended a few miles off the land, while the surface of the water towards the centre of the Firth was as smooth as glass. Numbers of Kittiwakes and Guillemots were scattered here and there in flocks as far as the eye could reach; often intermixed with these ravenous sea-fowl small parties of from two or three to half a dozen Shearwaters could be distinguished. The latter were engaged in busily swimming from one spot to another, occasionally skimming the water with their beaks, but rising on wing and gliding off when the Divers and Gulls took flight. A few single birds were met with, and these took little notice of the boat, seldom attempting to escape till within the distance of twenty or thirty yards; during the morning I procured two or three with perfectly white breasts, and one on which the plumage of the throat, breast, and belly was of a uniform dusky grey. No opportunity of examining the young having fallen to my share, I am unable to state the colouring of their first feathers; it is probable, however, that these dark-tinted birds were immature.

The flight of the Shearwater, when viewed for the first time, is sure to attract attention; and the bird as it glides with an undulating motion over the waves may readily be recognized at almost any distance. While cruising in company with the herring-fleet in the North Sea during the latter end of autumn, I observed at different times numbers of single birds, all apparently bound for the south, being usually met with in stormy or threatening weather, passing steadily on their course without turning to the right or left or pausing in their flight.

FULMAR.

PROCELLARIA GLACIALIS.

I MUCH regret that my knowledge of St. Kilda, derived from personal observation, consists merely of a somewhat hazy idea of the jagged outline of those rocky islets, viewed occasionally, as the sun disappeared in the Western Ocean, from the summit of one or two of the hills in the Outer Hebrides. Wind and weather, combined with other circumstances, having frustrated all my endeavours to reach the islands, I am unable to give the slightest information concerning the breeding-habits of this interesting species.

With the exception of a single specimen picked up in a disabled condition on the shore of the Firth of Forth in November 1863, I have met with the Fulmar only along the coast of Norfolk and in the North Sea. After the disastrous October gales in 1879 numbers were seen by the crews of the herring-boats and trawlers between twenty and thirty miles off the land. That their occurrence off this portion of the east coast is somewhat unusual, may be judged from the fact that the master of one of the luggers (an old gunner, and well acquainted with all sea-fowl) assured me that it was forty years since he had seen a Fulmar, when, as a "younker," he captured one in the "deidle." Many must have perished from the continued buffetings of the storms, as in addition to several reported to have been observed lying among the weed washed up on the shore, I met with two or three dead on the beach near Yarmouth Harbour, and noticed several others floating on the water outside the sands; in every instance these birds were dark in colouring, differing considerably from those seen on wing, the latter being invariably brightly tinted. The Fulmar found disabled on the shore of the Forth in 1863 was also exceedingly dusky, and proved to be the only one in that state of plumage that I met with alive. I had always imagined the dark birds to be immature, having almost doubted the assertions of a Highlander who informed me that when the young were fit to leave the nest they corresponded in colour with the adults. The statement, however, in the latest edition of Yarrell, that the young exhibiting down are white-breasted and similar in general tone of plumage to their parents, entirely upsets my conjecture, and even during the past year further information confirming this assertion has been received. As I am unable to form an opinion based on personal observation, I must leave to those who have met with better opportunities for examining the birds in a living state to decide whether the dark forms are merely a variety or a species. This same question appears to have puzzled most writers, as they tell us little on the subject.

On Sunday, the 2nd of November, 1879, I watched a light-plumaged bird flying along the shore at Yarmouth; sweeping round in large circles while holding a course towards the south, it occasionally passed over the sands as far as the drive, apparently utterly regardless of the traffic. On the 8th, another in similar plumage was observed, about thirty miles off the land, darting down among a swarm of Gulls collected round one of the luggers; some fish-liver having been thrown overboard from the steamboat, the patch of oil at once attracted its attention, and with a rapid swoop the stranger was over the spot.

Being anxious to secure the bird as a specimen, I did not waste time in watching its actions, but availed myself of the first chance. From the naturalist to whom it was sent for preservation I learned that the stomach contained, in addition to some fatty matter, a good handful of greasy oakum; and, strange to say, the manner in which the bird became possessed of this apparently unpalatable mouthful was afterwards ascertained. The previous day, while conversing with the master of one of the luggers (Henry Thomas, alias "Gaby" *), he promised, if we fell in with him while hauling his nets, to have the birds well fed with fish, by which means a larger number would be collected round the boat, and a greater chance afforded for observing or, if necessary, procuring specimens of Skuas or Fulmars. Though the night had been clear, the morning at daybreak proved hazy, and a drifting rain setting in shortly after, there was little chance of our making out any signals on boats at a distance. Before leaving the fleet it was ascertained from the crew of a Yarmouth lugger who had seen him shoot his nets that we were a few miles to the south of the station he had taken. Early in the following week I learned from Thomas that on the morning of the 8th he had seen no birds worth mentioning, with the exception of a few dark-plumaged Skuas (young Pomatorhines) and one Fulmar; the latter, he stated, came close under the stern of the boat, but though hovering round for a time, took no notice of the food thrown out. Happening, however, himself to fling overboard a piece of greasy oakum, with which he had been cleaning and oiling some of the gear, the bird immediately dashed down, and seizing the unsavoury morsel in its beak, sailed off towards the south. There could be little doubt that the specimen I obtained was the same bird: as near as we were able to judge, it was shot about half an hour after being lost sight of by my informant. While in correspondence with Mr. Gunn of Norwich, to whom the bird had been sent for preservation, I was furnished with the following notes referring to this specimen which he had taken down at the time:—"An adult female, exceedingly fat, weight 1 lb. 9 oz., length 18¼ inches, extent of wings when open 44 inches, the wing from carpal joint to tip of longest primary 12½ inches. Head pure white. The stomach contained a piece of oakum and some lumps of fat; the former had been apparently thrown overboard from a vessel, being saturated with grease, probably for machinery." Then referring to a dark-tinted bird picked up at sea, and forwarded to him the same day, he adds:—"The other was also a female, but a younger specimen." It is evident from this remark that Mr. Gunn entertained the idea I then held, viz. that the dark-coloured birds were the juveniles. During the early morning I had previously noticed a couple of light-plumaged Fulmars flying south; these passed together at the distance of two or three hundred yards, sailing round and round, without paying the slightest attention to the clouds of Gulls hovering after a boat on which the crew were shaking out and cleaning the nets. Shortly after daybreak a few dead Fulmars had been passed floating on the water, and while steaming back towards the harbour during the afternoon, we stopped and lowered the boat to pick up two or three lying in our course about twenty miles off the land. Though the shades of colouring varied, all exhibited the dark stage of plumage, and had evidently succumbed to the long-continued exposure to the force of the terrible gales that during the past month had swept over the North Sea, causing death from exhaustion to thousands of even the hardiest sea-fowl while on the passage to their winter-quarters. None but those who had been afloat off our eastern coasts or observed the waifs as well as the remnants of battered and decomposing carcasses cast up by the tide could credit the destruction dealt out to the feathered tribe. Since that date I have met with but one of this species; the bird was flying towards the north, some miles outside the Cross Sands, in January 1882, and paid not the slightest regard to some liver and fish-offal thrown out to draw its attention.

The colours of the soft parts of the white-breasted bird shot on the 8th of November 1879 were as follows:—Iris dark hazel. The hooked portion of the upper mandible lemon-yellow; the remainder lemon and greenish towards base; the ridge or tube at the base a dark horn. This protuberance falls away

* Referred to under the heading of the Lesser Black-backed Gull on page 3.

considerably after death; in a few years the various portions of the beak, with the exception of the points of both upper and lower mandibles, become somewhat contracted. Point of lower mandible lemon-yellow, the upper ridge lemon and the lower a pale grey. Legs, toes, and webs a pale silvery grey, the scales being finely marked out with a slightly darker shade; the nails white. The description of the beak of the dark grey bird found on the shore of the Firth of Forth, and noted down at the time, corresponds in almost every particular with those of the birds in the same plumage met with on the Norfolk coast:—Points of upper and lower mandibles pale yellow, remaining portions of both pale dull grey; the ridge or tube on the upper mandible a pale brownish horn, slightly darker in some specimens. A little pale pink flesh veined with red showing round the gape; inside of mouth a pale flesh. Legs, webs, and toes pale silvery grey; nails very pale horn. Iris dark hazel.

It is, I believe, generally allowed that the dark-coloured birds are smaller than those with white heads and breasts. An exceedingly dingy-coloured individual that I picked up, floating at sea, about twenty-five miles off the land, on November the 8th, 1879, is, however, considerably the largest specimen that has come under my notice, the beak being also especially bulky.

The figures in the Plate, showing the light and dark stages or forms, are taken from the birds obtained in the North Sea, off Yarmouth, on the 8th of November, 1879. As previously stated, the former was shot and the latter picked up while floating dead on the water.

STORMY PETREL.

PROCELLARIA PELAGICA.

This active little sea-bird may be seen (that is, if you can catch a glimpse of it), at one season or another, off every portion of our coast-line, round the whole of the British Islands. It is as well to state that I have had but few chances for making observations on their habits; a visit was paid, during the latter part of May 1868, to Fura, a small, rocky, uninhabited island off the west coast of Ross-shire, where they were declared to breed by the natives resident at the fishing villages on the mainland. The holes and the scratchings in the soil to which they resorted were examined, though not a bird could be found, and it was evident our exploration of their haunts had been undertaken a week or two before their arrival at their breeding-quarters. In the North Sea, while in company with the herring-fleets in autumn, I have often seen a few; and during, as well as after, the gale of November 1872 they were abundant off Great Yarmouth and all along the Norfolk coast. In the Channel, and especially off the shores of Sussex, they are exceedingly numerous during spring, and may be occasionally met with in autumn. This is all the information I can give concerning their distribution around our islands from personal observations, and few, if any, authors on ornithological subjects appear to be able to say much more from their own experience.

Several writers have stated that the appearance of the Stormy Petrel is, as its name would denote, an indication of bad weather and gales at sea; this is certainly a mistaken idea, as, with only a gentle breath of air, and the surface of the water almost as smooth as glass, the birds may be attracted to the spot if a small quantity of fish-liver is thrown overboard. The smell of the oil that then disperses through the air draws them from all quarters, to which it is carried in the space of a few minutes. I first heard of this manner of bringing the Petrel within range from a Brighton boatman, while fishing off the Black Rock, a favourite feeding-ground for the finny tribe, in the autumn of 1870; as we were busily engaged in hauling up the rock-whiting and other denizens of the deep*, a single bird was detected hovering round, though sheering off before the gun could be picked up and brought into use. The boatman declared that at times any number might be shot if some fish-liver was on board and a few small pieces flung out. Luckily a quantity happened to have been stowed away, and some small pieces having been dropped on the water, we anxiously awaited the result; shortly after, the bird was again flitting round, and, offering an easy chance, was procured.

During the spring, when this species is gathering in the Channel, and on the whole of our coast-line, before making a move to their northern breeding-stations, they are to be found almost every day that an attempt is made to ascertain their whereabouts.

* A three-bearded rockling, a rare and very beautifully marked and coloured fish, was taken that day, the first I ever caught, and no others were met with till I obtained a couple off Shoreham in September 1882—the first, captured on the 4th of the month, weighing 2 lbs., and the second a week later, bringing down the scales at 1¾ lb. Both of these fish were exhibited at the International Fisheries Exhibition in London, in 1883, by Mr. Gunn of Norwich, to whom I sent them, not being a collector of the finny tribe myself.

I have been enabled on several occasions, all of which have been entered in my notes, to make observations on this species, when attracted by the means previously referred to.

May 20th, 1872. Out to sea in a fishing-lugger off Brighton and Shoreham, in order to procure some Stormy Petrels. When about four miles from the land, we put over a quantity of skate's liver and oil *, and in a few minutes the first Petrel appeared upon the scene. We soon discovered that it was the best plan to throw out the liver in large pieces, as the birds are then unable to carry it off, and remain fluttering round while endeavouring to secure a mouthful now and then. Should the pieces be small, they speedily pick up as much food as they require and fly off without affording a chance for a shot. It is astonishing in how short a time, and from what a distance, they can be decoyed by the fish-liver on the water; in a little over half an hour I had shot eight, and having as many as were needed, we left the rest of the liver for the Gulls and Petrels, and proceeded to try the fishing at the wreck of the old lighter off Shoreham harbour. The Gulls, I remarked, were much more confiding than they had formerly been, owing, doubtless, to the freedom from persecution during the summer.

In order to protect the liver for the Petrel, I was obliged to shoot eight or ten Lesser Black-backed Gulls that were devouring all the liver as fast as we threw it out, flying almost on board our craft and snatching it up as soon as it fell on the water. There were very few fish this day about the remains of the wreck, and after bringing up for about four hours, we had only taken five score rock-whitings and about half a hundredweight of small congers. Several more Petrel came round the boat when the fish to be reserved for our own table were cleaned; rock-whiting lose all their flavour if not attended to immediately after they are captured.

1879, May 23rd. Out in the Channel off Brighton, in a steamboat, to look for Petrel, in hopes of falling in with the Forked-tailed Petrel (*Thalassidroma leachii*); only the smaller species were, however, observed, and these came readily to the liver and oil, followed in due course by numbers of Lesser Black-backed Gulls. These undesirable visitors we were forced to clear off, in order to preserve the food thrown out for the objects of our search.

24th. Dense fog in early morning, and very hot and still till the sun broke through. Again afloat soon after daybreak, but only fell in with the same species; plenty of Petrel, and too many Gulls flying round, requiring constant shooting to preserve our supply of fish-liver from being cleared off.

29th. In the afternoon my sixteen-foot boat, with a crew of two men, was towed out by a steamboat, about eight miles from the land, between Brighton and Shoreham, to have another turn at the Petrel. No sooner had I descended to our craft and cast off than a dense fog came drifting along from the east, obscuring the view beyond the distance of twenty or thirty yards. The captain of the steamboat at once stopped his vessel, and hailed us to ascertain if we had a compass, and being informed we were without one, went below and speedily returned on deck with the necessary article. Having promised to steam out for us after dark, did we not turn up by that time, he left us to follow our own inclination to remain in the small boat, which they all appeared to consider rather risky. Some liver and oil were put out on the water as the steamer moved off, and we waited a short time, but no Petrel were seen; after about half an hour's pull we halted again, though the fog was so dense that nothing beyond twenty yards could be discerned. Another lot of liver and oil was thrown overboard, and our rations were then served out; not a bird came in view while we consumed our allowance, and smoking had been commenced before we noticed that one of the oars had slipped overboard without previously attracting our attention. As we carried a spare set, no attempt was made to secure it at the time, but merely to keep it in sight; a few minutes later a Petrel was discovered settled on the blade, employed in pecking at a piece of liver which had drifted up against the oar. Shortly after, two or three more were hovering round, and when ready to move off, a couple were obtained with the two barrels, the rest speedily taking their departure. During

* This oil was obtained from the blubber of seals killed in the Dornoch Firth, off the east coast of Ross-shire, cut up into chunks and boiled down in a caldron.

our pull towards the land, through the dense fog, we stopped once more and threw over what remained of the cod-liver and oil. One or two again came round, but the drifting fog became so thick, that we could not see them at any distance, except too close to fire without damaging their plumage. The sound of traffic in the town caught our ears after another pull of half an hour, and the line of beach shortly after becoming visible, it was necessary to land to ascertain our whereabouts. It was satisfactory to learn that we were only a quarter of a mile to the west of the New Pier, for which I had attempted to steer. The crew of the steamboat, which was just being got ready to go in search of us, were much surprised to find we had reached our destination so early, after an eight miles' pull in the thick fog.

In 1880 I was again out in the steamboat, on the 13th of May, off Brighton, to ascertain if the Petrel were now passing along the coast. Starting at 5 P.M. from the New Pier, we steamed about seven miles south-west, and then put out the cod-liver and oil, and fifteen or sixteen birds were counted hovering round it during the time we remained near at hand. A small boat belonging to the steamer was lowered, as I had unfortunately omitted to order the one usually taken out, and it was only with considerable difficulty that I was enabled to shoot, as this dilapidated craft had no bottom boards on which to obtain a footing, and was utterly unfitted for going afloat. Notwithstanding these drawbacks, I managed to knock down and obtain five; double the number might, however, have been secured with ease had the boat been properly fitted, as the birds were but little disturbed by the report of the gun, and continued flitting round and pecking at the liver, and skimming over and dipping down to the oil without the slightest signs of alarm. Few Gulls were seen on this occasion, though several flocks of Common and Arctic Terns passed, and two or three large bodies of Knots. A young Gannet in the second year's plumage also came and circled round us while working away towards the east.

In November 1872, I was in the east of Norfolk, and the greater part of the month was passed at Yarmouth. On the 11th a terrible gale began to blow, with squalls of rain, from the north-north-east; large bodies of fowl were flying continually all day towards the north, and the immense flocks of Dunlins, passing in rapid succession, were such as I had never witnessed before or since.

On the 12th the gale was from the east-north-east; a few fowl were still flying north, but the number of Dunlins seen on wing were not to be compared with those observed the previous day.

Still blowing with fearful gusts from the east on the 13th; wild fowl in small numbers were now and then detected skimming over the breakers, on their way towards the north, but the Dunlins appeared to have all passed, not one being observed during the whole of the day. There were hundreds of Gulls fishing in the harbour tide, where it ran into the roads, but only the usual visitors to the spot were observed, with the exception of a single immature Glaucous Gull. Three or four Stormy Petrel remained feeding just off the harbour-mouth, and occasionally flying a short distance between the piers, where they were enabled to find shelter from the cutting blasts.

On the 14th blowing still harder from the east-north-east, with frequent squalls of hail and snow. Several vessels ashore between Yarmouth and Lowestoft, and much wreckage observed floating in the roads. There was a large flock of Gulls off the harbour and a few flying up the river; some Kittiwakes were blown on to the denes, and several were with the swarms of Grey Gulls in the roads, evidently much cut up by the rough usage they had suffered during the long continuation of the gale. A dozen or so of Stormy Petrel were flitting here and there between the piers near the harbour-mouth, and a few made their way up the river as far as the wharfs where the vessels and fishing-boats were moored.

On the 15th there was a slight lull in the morning, and all the common species of Gulls were seen in hundreds just off the harbour-mouth during the whole of the day, several Stormy Petrel flying with them.

At daybreak on the 16th the wind had freshened, and a gale was blowing hard from the east-south-east. The unfortunate little Petrel proved to be much exhausted by the long continuation of the storms, and were even driven right up the harbour, several being seen on Dreydon mudflats. Off the denes to the north of the harbour they were hovering in dozens over the rolling breakers, and were frequently carried by the irresistible squalls of wind right across the carriage-drive. I once remarked an immature Pomatorhine Skua attempting to hold to windward over the beach in front of the town, but the weary stranger was more than once carried back by the drifting squalls, and seemed in an extremely weak and helpless condition.

The surface of the North Sea on Monday morning, the 18th, was nearly as smooth as glass, the wind having completely dropped; and leaving the harbour early in the 'Reliance,' we steamed out through the roads and round the sands. Immense numbers of Stormy Petrel were resting on the water, most of them asleep, with their heads buried under the feathers of their backs, strongly resembling small round balls as they floated about, with their plumage puffed out, in the tideway. They were passed during the whole of the day along the coast from Winterton to Lowestoft, many hundreds being noticed. No Forked-tailed Petrels were observed, though I carefully examined with the glasses all that were within view, as we steamed slowly past them. These poor little birds had evidently been completely tired out by the long continuation of the gale, and were now endeavouring to recruit their strength by seeking repose on the placid surface of the water.

On the 20th I was out to the Newarp lightship, and then round the Cross Sands returning by St. Nicholas Gat. We boarded the Newarp, and proved to be the first vessel spoken since the gale, which had been ridden with one hundred and sixty fathom of cable; the crew had naturally had a rough time of it, still everything on board was in the greatest possible order and cleanliness. The only birds that had come on board during the week of the gale were two Stormy Petrel with a few of our Common Finches and a female Blackbird.

The Stormy Petrel, so numerous close in shore two days previous, had now almost entirely disappeared, and only two or three were noticed flitting about in an apparently worn-out condition outside the Cross Sands.

During May 1881, many Stormy Petrel were reported as seen in the Channel off Brighton and Shoreham by fishermen and shore-shooters, and some of the Forked-tailed species were also declared to have been killed; one of the latter I, however, discovered to be a Nightjar, shot a couple of miles at sea while on the passage to its breeding-haunts. Such ridiculous mistakes are unfortunately of not unfrequent occurrence, and doubtless at times lead to false entries of rare birds being recorded in works on natural history.

The earliest mention in my notes concerning the Stormy Petrel is made in December 1864, when a bird of this species was observed just before dusk to strike the front of one of the large houses in the Marina, a row facing the Channel, at St. Leonards-on-Sea, in Sussex, and came fluttering down into the area. This occurred shortly after a heavy gale, and I came to the conclusion that the poor little mite must have been so weary and worn out by the continued exposure to the force of the wind and squalls of rain that had lately swept over this part of the coast, as to be utterly oblivious as to the course it was following.

While referring to this species, in vol. vi. of his 'History of British Birds,' the Rev. F. O. Morris remarks: "It has received its name of Petrel from its habit of walking or running on the surface of the water, as the Apostle St. Peter did, or essayed to do." I could never understand why these birds were called "Mother Carey's Chickens;" and in hopes of ascertaining the reason for this appellation, an inquiry was made to the very obliging editor of 'The Field,' and the following answer appeared on the 16th of January, 1886:—

"E. T. B.—It is supposed that the name 'Mother Carey' is a corruption of *Madre cara* (dear mother), addressed by pious seamen to the Virgin when beseeching her aid to avert a storm."

ROUGH NOTES

ON THE

BIRDS OBSERVED

DURING TWENTY YEARS' SHOOTING AND COLLECTING

IN THE

BRITISH ISLANDS.

BY

E. T. BOOTH.

WITH PLATES FROM DRAWINGS BY E. NEALE

TAKEN FROM SPECIMENS IN THE AUTHOR'S POSSESSION.

PART I.

Golden Eagle. (4 Plates.)	Osprey. (1 Plate.)
White-Tailed Eagle. (1 Plate.)	Kite. (2 Plates.)

LONDON:

PUBLISHED BY R. H. PORTER, 6 TENTERDEN STREET, W

AND

MESSRS. DULAU & CO., SOHO SQUARE, W

1881.

PRINTED BY TAYLOR AND FRANCIS,
RED LION COURT, FLEET STREET.

ROUGH NOTES

ON THE

BIRDS OBSERVED

DURING TWENTY YEARS' SHOOTING AND COLLECTING

IN THE

BRITISH ISLANDS

BY

E. T. BOOTH.

WITH PLATES FROM DRAWINGS BY E. NEALE,

TAKEN FROM SPECIMENS IN THE AUTHOR'S POSSESSION.

PART II.

Common Buzzard. (2 Plates.)
Peregrine Falcon. (1 Plate.)
Merlin. (1 Plate.)
Kestrel.
Sparrow-Hawk. (1 Plate.)
Marsh-Harrier.
Hen-Harrier. (1 Plate.)
Short-eared Owl. (2 Plates.)
Long-eared Owl. (1 Plate.)
Tawny Owl.
Barn-Owl. (1 Plate.)
Red-backed Shrike.
Great Tit.
Coal Tit.
Crested Tit. (1 Plate.)
Blue Tit.
Marsh-Tit.
Long-tailed Tit.
Bearded Tit (2 Plates.)
Pied Flycatcher.
Spotted Flycatcher.
Kingfisher.

LONDON:
PUBLISHED BY R. H. PORTER, 6 TENTERDEN STREET, W.,
AND
MESSRS. DULAU & CO., SOHO SQUARE, W.
1882.

PRINTED BY TAYLOR AND FRANCIS,
RED LION COURT, FLEET STREET.

ROUGH NOTES

ON THE

BIRDS OBSERVED

DURING TWENTY YEARS' SHOOTING AND COLLECTING

IN THE

BRITISH ISLANDS

BY

E. T. BOOTH.

WITH PLATES FROM DRAWINGS BY E. NEALE,

TAKEN FROM SPECIMENS IN THE AUTHOR'S POSSESSION.

PART III.

Raven. (1 Plate.)	Nuthatch.	Swift.
Black Crow.	Wryneck.	Swallow.
Grey Crow. (1 Plate.)	Creeper.	House-Martin.
Jackdaw.	Green Woodpecker.	Sand-Martin.
Chough.	Great Spotted Woodpecker.	Crossbill. (1 Plate.)
Magpie.	Lesser Spotted Woodpecker.	Wren.
Jay.	Nightjar. (1 Plate.)	

LONDON:

PUBLISHED BY R. H. PORTER, 6 TENTERDEN STREET, W.,

AND

MESSRS. DULAU & CO., SOHO SQUARE, W.

1882.

PRINTED BY TAYLOR AND FRANCIS,
RED LION COURT, FLEET STREET.

ROUGH NOTES

ON THE

BIRDS OBSERVED

DURING TWENTY YEARS' SHOOTING AND COLLECTING

IN THE

BRITISH ISLANDS

BY

E. T. BOOTH.

WITH PLATES FROM DRAWINGS BY E. NEALE,

TAKEN FROM SPECIMENS IN THE AUTHOR'S POSSESSION.

PART IV.

Pied Wagtail.
White Wagtail.
Grey Wagtail.
Grey-headed Wagtail. (1 Plate.)
Yellow Wagtail. (1 Plate.)
Meadow-Pipit.
Tree-Pipit.
Rock-Pipit. (1 Plate.)
Redstart.
Black Redstart. (1 Plate.)
Stonechat.
Whinchat.
Wheatear. (1 Plate.)
Grasshopper Warbler.
Garden-Warbler.
Whitethroat.
Lesser Whitethroat.
Wood-Wren.
Willow-Wren. (1 Plate.)
Chiffchaff.

LONDON:

PUBLISHED BY R. H. PORTER, 6 TENTERDEN STREET, W.

AND

MESSRS. DULAU & CO., SOHO SQUARE, W.

1883.

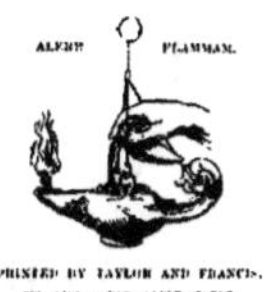

PRINTED BY TAYLOR AND FRANCIS,
RED LION COURT, FLEET STREET.

ROUGH NOTES

ON THE

BIRDS OBSERVED

DURING TWENTY YEARS' SHOOTING AND COLLECTING

IN THE

BRITISH ISLANDS

BY

E. T. BOOTH.

WITH PLATES FROM DRAWINGS BY E. NEALE,
TAKEN FROM SPECIMENS IN THE AUTHOR'S POSSESSION.

PART V.

Shore-Lark.
Wood-Lark.
Bunting.
Black-headed Bunting.
Yellow Bunting.

Cirl-Bunting.
Chaffinch.
Brambling.
Sedge-Warbler.
Reed-Warbler.

Nightingale.
Curlew.
Red-breasted Merganser. (2 Plates.)
Gannet. (6 Plates.)

LONDON:
PUBLISHED BY R. H. PORTER, 6 TENTERDEN STREET, W.,
AND
MESSRS. DULAU & CO., SOHO SQUARE, W.
1883.

PRINTED BY TAYLOR AND FRANCIS,
RED LION COURT, FLEET STREET.

ROUGH NOTES

ON THE

BIRDS OBSERVED

DURING TWENTY YEARS' SHOOTING AND COLLECTING

IN THE

BRITISH ISLANDS

BY

E. T. BOOTH.

WITH PLATES FROM DRAWINGS BY E. NEALE,

TAKEN FROM SPECIMENS IN THE AUTHOR'S POSSESSION.

PART VI.

Starling.	Blackbird.	Redshank.
Dipper.	Blackcap.	Purple Sandpiper.
Fieldfare.	Ptarmigan. (3 Plates.)	Coot.
Thrush.	Peewit.	Arctic Skua. (4 Plates.)

LONDON:

PUBLISHED BY R. H. PORTER, 6 TENTERDEN STREET, W.,

AND

MESSRS. DULAU & CO., SOHO SQUARE, W.

1884.

PRINTED BY TAYLOR AND FRANCIS,
RED LION COURT, FLEET STREET.

ROUGH NOTES

ON THE

BIRDS OBSERVED

DURING TWENTY YEARS' SHOOTING AND COLLECTING

IN THE

BRITISH ISLANDS

BY

E. T. BOOTH.

WITH PLATES FROM DRAWINGS BY E. NEALE,

TAKEN FROM SPECIMENS IN THE AUTHOR'S POSSESSION.

PART VII.

Rook.
Tree-Sparrow.
Redwing.
Golden-crested Wren.

Whimbrel.
Eider. (3 Plates.)
Goosander. (2 Plates.)

Common Gull.
Herring-Gull.
Pomatorhine Skua. (3 Plates.)

LONDON:
PUBLISHED BY R. H. PORTER, 6 TENTERDEN STREET, W.,
AND
MESSRS. DULAU & CO., SOHO SQUARE, W.
1884.

PRINTED BY TAYLOR AND FRANCIS,
RED LION COURT, FLEET STREET.

ROUGH NOTES

ON THE

BIRDS OBSERVED

DURING TWENTY YEARS' SHOOTING AND COLLECTING

IN THE

BRITISH ISLANDS

BY

E. T. BOOTH.

WITH PLATES FROM DRAWINGS BY E. NEALE,

TAKEN FROM SPECIMENS IN THE AUTHOR'S POSSESSION.

PART VIII.

✓ Montagu's Harrier. (2 Plates.)
Greenfinch.
Twite.
✓ Missel-Thrush.
✓ Quail.
✓ Woodcock. (2 Plates.)
✓ Jack Snipe.
Land-Rail.
Spotted Crake.
✓ Brent Goose.
✓ Whooper. (1 Plate.)
Pochard. (1 Plate.)
Goldeneye. (1 Plate.)
Long-tailed Skua. (1 Plate.)

LONDON:
PUBLISHED BY R. H. PORTER, 6 TENTERDEN STREET, W.,
AND
MESSRS. DULAU & CO., SOHO SQUARE, W.
1884.

PRINTED BY TAYLOR AND FRANCIS,
RED LION COURT, FLEET STREET.

ROUGH NOTES

ON THE

BIRDS OBSERVED

DURING TWENTY YEARS' SHOOTING AND COLLECTING

IN THE

BRITISH ISLANDS

BY

E. T. BOOTH.

WITH PLATES FROM DRAWINGS BY E. NEALE.
TAKEN FROM SPECIMENS IN THE AUTHOR'S POSSESSION.

PART IX.

Cuckoo.
Lark.
Hedge-Sparrow.
Dartford Warbler.
Wood-Pigeon.
Turtle-Dove.
Dotterel. (1 Plate.)
Black-tailed Godwit. (1 Plate.)
Water-Rail.
Moorhen.
Shoveller. (3 Plates.)
Great Crested Grebe. (1 Plate.)
Lesser Black-backed Gull.
Great Black-backed Gull. (1 Plate.)
Fulmar. (1 Plate.)

LONDON:
PUBLISHED BY R. H. PORTER, 6 TENTERDEN STREET, W.,
AND
MESSRS. DULAU & CO., SOHO SQUARE, W.
1885.

PRINTED BY TAYLOR AND FRANCIS,
RED LION COURT, FLEET STREET.

ROUGH NOTES

ON THE

BIRDS OBSERVED

DURING TWENTY YEARS' SHOOTING AND COLLECTING

IN THE

BRITISH ISLANDS

BY

E. T. BOOTH.

WITH PLATES FROM DRAWINGS BY E. NEALE,

TAKEN FROM SPECIMENS IN THE AUTHOR'S POSSESSION.

PART X.

House-Sparrow. (2 Plates.)
Bullfinch.
Redbreast.
Black Grouse. (1 Plate.)
Red Grouse. (1 Plate.)
Spoonbill. (1 Plate.)
Grey-lag Goose. (1 Plate.)
White-fronted Goose.
Scoter. (1 Plate.)
Smew. (1 Plate.)
Sclavonian Grebe.
Little Grebe.

LONDON:
PUBLISHED BY R. H. PORTER, 6 TENTERDEN STREET, W.,
AND
MESSRS. DULAU & CO., SOHO SQUARE, W.
1886.

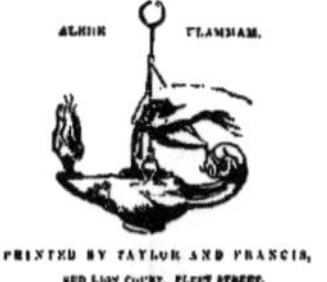

PRINTED BY TAYLOR AND FRANCIS,
RED LION COURT, FLEET STREET.

ROUGH NOTES

ON THE

BIRDS OBSERVED

DURING TWENTY YEARS' SHOOTING AND COLLECTING

IN THE

BRITISH ISLANDS

BY

E. T. BOOTH.

WITH PLATES FROM DRAWINGS BY E. NEALE,

TAKEN FROM SPECIMENS IN THE AUTHOR'S POSSESSION.

PART XI.

Ring-Ouzel.
✓ Rock-Dove. (1 Plate.)
✓ White Stork.
✓ Turnstone.
✓ Spotted Redshank.
✓ Snipe. (1 Plate.)
✓ Curlew Sandpiper. (1 Plate.)
✓ Knot. (1 Plate and a Map of Breydon Mudflats.)
Scaup.
Puffin. (1 Plate.)
Razorbill.
Cormorant.
Roseate Tern. (1 Plate.)
Arctic Tern.
Little Gull. (1 Plate.)

LONDON:
PUBLISHED BY R. H. PORTER, 6 TENTERDEN STREET, W.,
AND
MESSRS. DULAU & CO., SOHO SQUARE, W.
1886.

PRINTED BY TAYLOR AND FRANCIS,

ROUGH NOTES

ON THE

BIRDS OBSERVED

DURING TWENTY YEARS' SHOOTING AND COLLECTING

IN THE

BRITISH ISLANDS

BY

E. T. BOOTH.

WITH PLATES FROM DRAWINGS BY E. NEALE,

TAKEN FROM SPECIMENS IN THE AUTHOR'S POSSESSION.

PART XII.

Siskin. (1 Plate.)
Linnet.
Lesser Redpoll.
Mealy Redpoll.
Stock-Dove.
Ringed Plover.
Kentish Plover.
Sanderling.
Oyster-Catcher.

Greenshank.
Common Sandpiper. (1 Plate.)
Bar-tailed Godwit.
Little Stint. (1 Plate.)
Temminck's Stint. (2 Plates.)
Bean-Goose.
Mute Swan. (Map of Hickling Broad.)

Velvet Scoter.
Eared Grebe.
Black-throated Diver.
Red-throated Diver.
Black Guillemot
Sandwich Tern. (1 Plate.)
Manx Shearwater.

LONDON:
PUBLISHED BY R. H. PORTER, 6 TENTERDEN STREET, W.
AND
MESSRS. DULAU & CO., SOHO SQUARE, W.
1886.

PRINTED BY TAYLOR AND FRANCIS,
RED LION COURT, FLEET STREET.

ROUGH NOTES

ON THE

BIRDS OBSERVED

DURING TWENTY YEARS' SHOOTING AND COLLECTING

IN THE

BRITISH ISLANDS

BY

E. T. BOOTH.

WITH PLATES FROM DRAWINGS BY E. NEALE,

TAKEN FROM SPECIMENS IN THE AUTHOR'S POSSESSION.

PART XIII.

LONDON:

PUBLISHED BY R. H. PORTER, 6 TENTERDEN STREET, W.

AND

MESSRS. DULAU & CO., SOHO SQUARE, W.

1886.

PRINTED BY TAYLOR AND FRANCIS,
RED LION COURT, FLEET STREET.

ROUGH NOTES

ON THE

BIRDS OBSERVED

DURING TWENTY YEARS' SHOOTING AND COLLECTING

IN THE

BRITISH ISLANDS

BY

E. T. BOOTH.

WITH PLATES FROM DRAWINGS BY E. NEALE,

TAKEN FROM SPECIMENS IN THE AUTHOR'S POSSESSION.

PART XIV.

Goldfinch.
Hawfinch.
Shag.
✓Bewick's Swan.
Tufted Duck.
Long-tailed Duck.

✓Golden Plover.
✓Grey Plover.
Grey Phalarope. (1 Plate.)
Lesser Tern.
White-winged Black Tern.
Black Tern.

Kittiwake. (1 Plate.)
Glaucous Gull. (1 Plate.)
Black-headed Gull. (2 Plates.)
Stormy Petrel. (1 Plate.)
Great Northern Diver.
Common Guillemot. (2 Plates.)

LONDON:
PUBLISHED BY R. H. PORTER, 6 TENTERDEN STREET, W.,
AND
MESSRS. DULAU & CO., SOHO SQUARE, W.
1887.

PRINTED BY TAYLOR AND FRANCIS,
RED LION COURT, FLEET STREET.

ROUGH NOTES

ON THE

BIRDS OBSERVED

DURING TWENTY YEARS' SHOOTING AND COLLECTING

IN THE

BRITISH ISLANDS

BY

E. T. BOOTH.

WITH PLATES FROM DRAWINGS BY E. NEALE,

TAKEN FROM SPECIMENS IN THE AUTHOR'S POSSESSION.

PART XV.

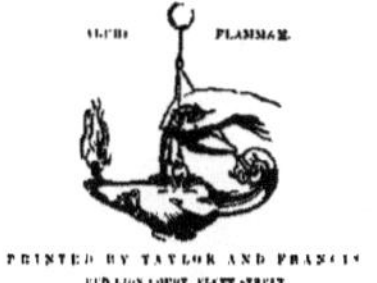

PRINTED BY TAYLOR AND FRANCIS
RED LION COURT, FLEET STREET.

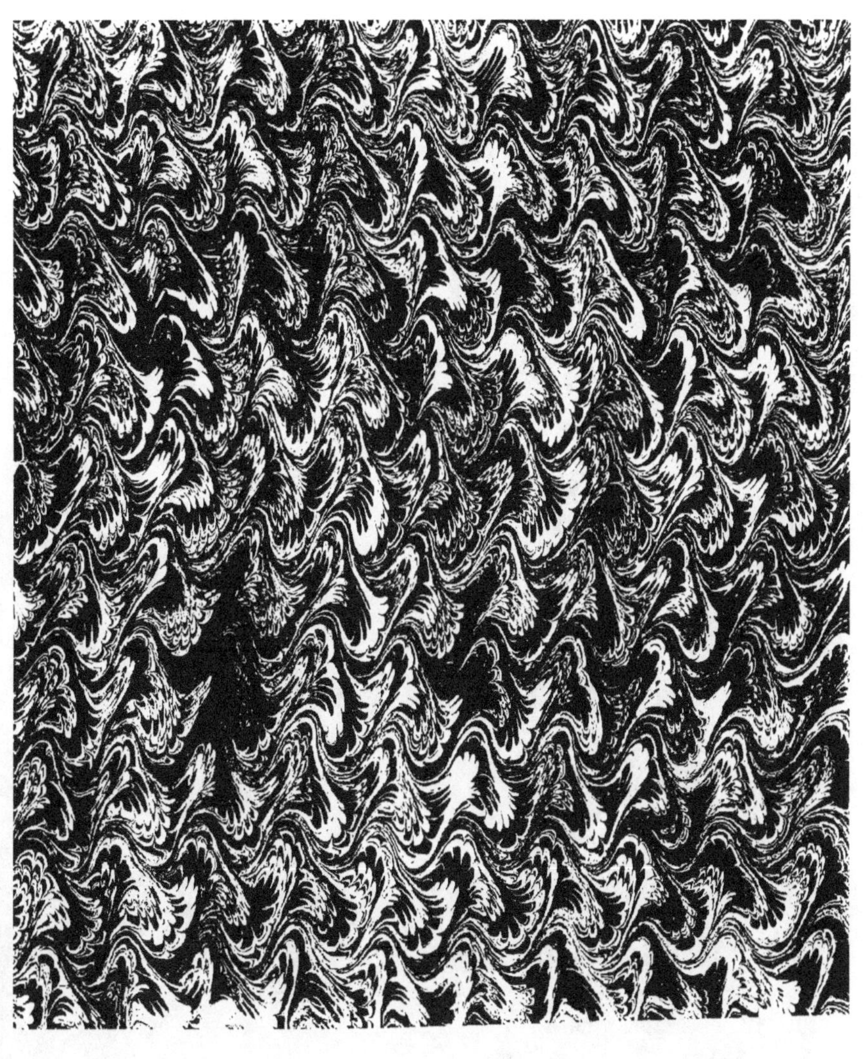

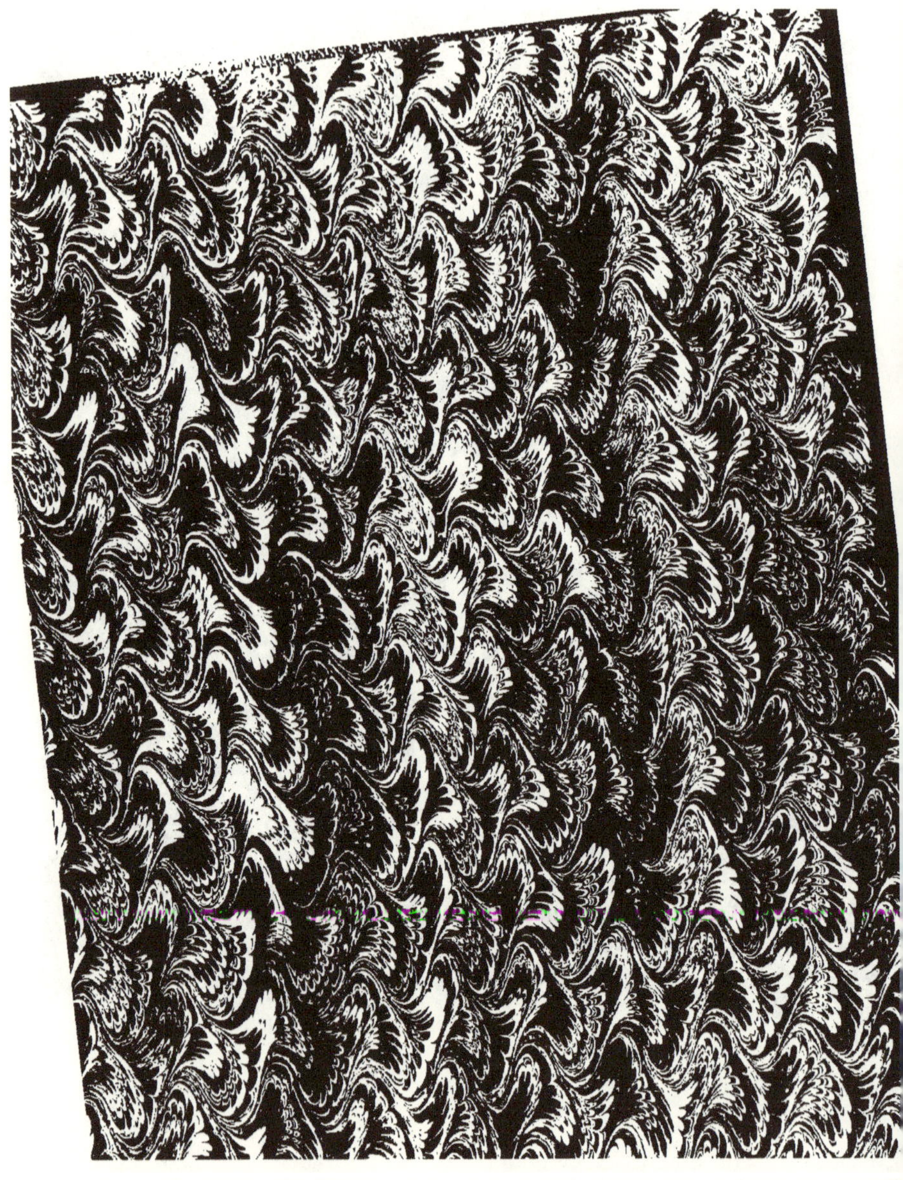

www.ingramcontent.com/pod-product-compliance
Lightning Source LLC
Chambersburg PA
CBHW021941220326
41599CB00011BA/973
9783742862839